OUTCROPPINGS

OUTCROPPINGS

JOHN McPHEE

Photographs by

TOM TILL

Edited by

CHRISTOPHER MERRILL

PEREGRINE SMITH BOOKS
SALT LAKE CITY

First edition

90 89 88   5 4 3 2 1

Introduction copyright © 1988 by John
McPhee. Selections excerpted from *Basin
and Range, Rising from the Plains,* and
Encounters with the Archdruid copyright ©
1971, 1981, and 1986 by John McPhee.
All rights reserved. Reprinted by arrange-
ment with Farrar, Straus & Giroux, Inc.,
19 Union Square West, NY, NY 10003.

Photographs copyright © Tom Till.
All rights reserved.

Copyright © 1988 by Gibbs Smith,
Publisher

Published by Gibbs Smith, Publisher,
P.O. Box 667, Layton, UT 84041

Design by J. Scott Knudsen

Printed and bound in Korea

All cover photographs copyright
© Tom Till
Front Cover:
    "Fossil," in the Ulrich Gallery near
    Fossil Butte National Monument
    Inset, "Canyonlands at Sunset"
Back Cover:
    "Grand Prismatic Spring"

**Library of Congress Cataloging-in-
Publication Data**
McPhee, John A.
     Outcroppings / John McPhee;
photographs by Tom Till.
         p.   cm.
     Excerpts from Encounters with the
archdruid, Basin and range, and Rising
from the plains.
     ISBN 0-87905-262-7:
     1. West (U.S.)—Description and
travel. 2. West (U.S.)—Historical
geography. 3. Geology—West (U.S.)
4. Physical I. Till, Tom. II. Title.
F591.M39 1988
917.8—dc19

To the memory of
Ethel Waxham Love
1882-1959

CONTENTS

PREFACE

P oetry is what gets lost in translation," Robert Frost wrote, admonishing for all time those who labor in the clumsy art of rendering one thing into another. And the editor of a book such as *Outcroppings* must also feel the poet's strict regard; for this work, too, is a translation, in the same sense that a film may be a "translation" of a novel. In fact, selecting passages from John McPhee's *Encounters with the Archdruid, Basin and Range,* and *Rising from the Plains* to accompany a photographic essay posed for me some of the problems a screenwriter faces in turning a novel into the script for a film: what to leave in, what to leave out, how to tell the story quickly.

Editing this book, then, was at once exhilarating and impossible. McPhee's stylish prose; his gift for breathing life into the most disparate subjects; his wide range of characters; his deep respect for landscape; his sly wit and generosity;—these made my task exhilarating. Impossible as well, because there was too much to choose from. Selection necessitates exclusion. What's missing here is the delicate web of argument spun through each of his books, the line of reasoning which will reveal its many beauties, pleasures, and complexities only over the course of a longer narrative—in short, some of the poetry.

I offer instead a single journey distilled from McPhee's various wanderings around the West, recognizing that it is, of course, simply a version of my own journey through his work. A map of my reading, so to speak, the central features of which include westward migration;

homesteading; the landscape's splendors, resources, and dangers, as well as our shifting relationship with it. No doubt other journeys are possible; great books sustain a host of different readings.

And they may well leave their readers with more questions than answers.

How, for example, does a man from Princeton, New Jersey, who works within walking distance of the very hospital in which he was born, come to write about the geology of the West? I can only speculate. And as for what effects such a decision might have on his writings, his vision of the world? Or where his explorations might lead him next? "There are limits to imagination," the poet Robert Hass reminds us.

A word about the photographs: they form a narrative designed to stand on its own, a text which runs parallel to the written part of the book, now approaching a descriptive passage in documentary fashion, now veering off into a stylized statement. In no way are the photographs meant to "explain" John McPhee (as if he needs explanation!); nor should his writings amplify the photographs. Ideally, the two texts work in counterpoint, creating sparks across the page—poetry, if you will—to replace the poetry sacrificed in the initial editing.

Christopher Merrill
Santa Fe, New Mexico

"In the vernacular of geology, your nose was on the outcrop. Through experience with structure, you reached for the implied tectonics. Gradually, as you gathered a piece here, a piece there, the pieces framed a story."

Rising from the Plains

INTRODUCTION

To the editor's question — what prompted me to write about the West? — the answer is: a tennis match at Forest Hills. Arthur Ashe was playing Clark Graebner in the semifinals of the first National Open Tennis Tournament, in September 1968. Both were Americans. Both were twenty-five. Therefore, they would have known each other for at least fourteen years, because players of that calibre are so rare that almost surely they will meet as children.

For ten years or so, I had been writing journalistic sketches of individuals: actor, forager, art historian, writer, headmaster, basketball player — one at a time. The thought had often come to mind that if I were to sketch a pair of people in a single piece of writing — a pair whose idiosyncrasies and life histories would be interreflective — a result might be that one plus one would add up to more than two. Architect and client, actor and director, pitcher and manager, choreographer and dancer — for several years I had mulled the idea without doing much about making a choice. Two or three weeks after the Ashe-Graebner match, something stirred, and I realized that they were just the sort of pair I'd been seeking. After all, their physical skills were so considerable that the essential differences between them must largely be in their minds, and such things would emerge in a narrative of the match. An idea that had been half asleep was suddenly in a state of frenzy. CBS had broadcast the match and I would need a copy of the tape as the basic notebook for what I meant to do. I picked up the phone. CBS said I had not called a minute too soon. The tape was scheduled for erasure that afternoon.

The story, called "Levels of the Game," worked out

all right, and therefore caused me, looking forward, to raise the ante. I tacked to a wall a piece of paper that presented as a large enigmatic fraction the letters ABC over D. The A and the B and the C were meant to be people, each separately presented in a sort of Ashe-Graebner relationship with D. At that point, I had no idea what the subject would be, let alone what people would be involved. This is no way to go about developing a piece of writing—to start with a structural concept and then look around for a theme. I had not done it before and have not done it since, but it put me under a backpack in the North Cascades and at water level through the Grand Canyon. It took me to twelve thousand feet in the Sierra Nevada and into the foothills of the Klamath Mountains in the overshadowing dusk of big trees. As it happened, I had been to San Francisco before then, and also to Los Angeles, but I had never been in the West.

Another general question about the choice of a subject for a piece of writing is, Why choose that one over all other concurrent possibilities? Why does someone whose interest is to write about real people and real places choose these people, those places? For such projects, ideas are everywhere. They just stream by like neutrons. Since it may take a month, ten months, or three years to turn one idea into a piece of writing, what governs the choice? Not long ago, I made a list of all the pieces I had written in the past twenty years, and then put a check mark beside each one whose subject related to things I had been interested in before I went to college. I checked off more than ninety per cent.

My father was a medical doctor who dealt with the injuries of Princeton University athletes, travelled the world with United States Olympic teams, and, when I was very young, spent summers as physician at a Vermont camp. The place was called Keewaydin and was a classroom of the woods. It specialized in canoe trips and backpacking trips and taught ecology in our modern sense when the word was still connoting the root and shoot relations of communal plants. I grew up there, summers, ages six to twenty, ending up as a leader of those trips.

I played basketball and tennis there, and on my high school teams at home, with absolutely no idea that I was building the shells of future pieces of writing. I dreamed all year of the trips in the wild, not imagining, of course, that they would eventually lead to the Brooks Range, to the Yukon-Tanana suspect terrain, to the shiplike ridges of Nevada and the Laramide mountains of Wyoming. They would lead to the rapids of the Grand Canyon in the company of C over D.

Those letters went up on the wall in the early days of the ecologic uplift, when the environmental movement was gathering strength. That seemed to be an obvious choice for a theme. D would be an outspoken conservationist and the three others his natural enemies. With John Kauffmann, a friend of mine who was then a planner in the National Park Service, I made a list of six or seven D's and seventeen possibilities for A, B, and C: dam builders, developers, economic geologists—not simple environmental vandals, but people who believed in their form of husbandry of the earth and its resources. D, ultimately, was David Brower, the first executive director of the Sierra Club, who seemed the best choice not because he was the most reasonable conservationist in the United States but precisely because he was not. In the field, he appeared to have the sharpest and most active teeth.

I don't think I need to dwell on the appeal of the montane West to a person from New Jersey. Once I had been down the Colorado and up on high ground with Brower and the others, I knew I would never be of that world but again and again would be in it. I didn't know then how very much that would include Alaska, or how caught up I would become in the architectonics of the landscape itself. When I was seventeen, I went to Deerfield Academy, in Massachusetts, where a geologist named Frank Conklin presented his subject in a first-rate full-year course. Even then, I was an English-major designate, but in the years of writing that followed the geology always lay there to be tapped. Sooner or later in most of my writing projects, geology would be touched upon in one way or another, and I would ask the geologists of the Princeton faculty to help me get it right. There were some geological passages in *Encounters with the Archdruid*—ABC over G—and there were more in *Coming into the Country,* arising from a question I had long meant to ask. Obviously, the placer gold in the drainages of the Yukon was there because weather had broken up mountains and bestrewn the gold in the gravels of streams. That I thought I understood. But I wondered what had put the gold in the mountains in the first place. I called the geology department and talked with a professor who said he could not begin to answer the question. He had a much greater interest in Jurassic leaves. "Call Ken Deffeyes," he said. "Deffeyes knows, or thinks he knows." For me, Deffeyes put the gold in the mountains.

A year or so later, in an idle conversation with this same eclectic petrologue, I asked if he thought that we might find a Talk of the Town piece for the *New Yorker* magazine in a roadcut near the city. We could look at the blast-exposed face of the rock, read its history, and tell it in the first-person plural. Deffeyes said he would be pleased to be interviewed for the Talk piece, and thought such a thing might indeed work out. While we were still planning this short trip, I asked him if there would not be an even better story in a journey north from roadcut to roadcut—for example, up the Northway through the Adirondacks.

"Not on this continent," said Deffeyes. "If you want to do that sort of thing on this continent, go west—go across the structure."

In one moment, vaulting and rash, my thoughts raced to San Francisco with roadcuts lining the route like billboards, each with its own message. "Why not go all the way?" I said to him. Two weeks later, we were looking for silver in Nevada.

With that, I began travelling back and forth across the continent—on and off Interstate 80—with various geologists. My purpose was to describe not only the rock but the geologists as well, in a series of pieces with the overall title "Annals of the Former World." The result was meant to be a sort of cross-section of the United States along the fortieth parallel, and a picture of the science as it was settling down with its revolutionary theory of plate tectonics. That was the controlling element—the theme that would guide the entire composition. So the structure is not linear—not a straightforward trip from New York to San Francisco. It jumps about the country. It begins in New Jersey and leaps to Nevada because the tectonics in New Jersey two hundred million years ago are being recapitulated by the tectonics in Nevada today.

After *Basin and Range,* which presents the theory, and *In Suspect Terrain,* in which Anita Harris attacks it, the series moved to Wyoming and *Rising from the Plains.* The book is about a Rocky Mountain regional geologist named David Love, who was born in the center of Wyoming in 1913 and grew up on an isolated ranch, where he was educated mainly by his mother. To be sure, experience had come to him beyond Wyoming's borders—a Yale Ph.D., explorations for oil in the southern Appalachians—but his career had been accomplished almost wholly in his home terrain, and for several decades he had been regarded by colleagues as one of the two or three most influential field geologists in the United States Geological Survey. Inevitably, he had come to be called "the grand old man of Rocky Mountain geology."

After I had known him for a couple of years and made a number of field trips in his company, he reached into a drawer in his office in Laramie and handed me a journal that had been started by his mother long before she was married—a slim young woman who stepped down from a train in Rawlins, Wyoming, to go north by stagecoach into country that was still very much the Old West. When her son gave me her journal, she would have been over a hundred years old, and needless to say I never met her, but the admiration and affection I came to feel toward her is probably matched by no one I've encountered in my professional life. To the editor's question—why would this writer be drawn to the West?—she and her son are enough of an answer.

1.   WESTWARD

GEOLOGISTS ON THE WHOLE are inconsistent drivers. When a roadcut presents itself, they tend to lurch and weave. To them, the roadcut is a portal, a fragment of a regional story, a proscenium arch that leads their imaginations into the earth and through the surrounding terrain. In the rock itself are the essential clues to the scenes in which the rock began to form—a lake in Wyoming, about as large as Huron; a shallow ocean reaching westward from Washington Crossing; big rivers that rose in Nevada and fell through California to the sea. Unfortunately, highway departments tend to obscure such scenes. They scatter seed wherever they think it will grow. They "hair everything over"—as geologists around the country will typically complain.

"We think rocks are beautiful. Highway departments think rocks are obscene."

"In the North it's vetch."

"In the South it's the god-damned kudzu. You need a howitzer to blast through it. It covers the mountainsides, too."

"Almost all our stops on field trips are at roadcuts. In areas where structure is not well exposed, roadcuts are essential to do geology."

"Without some roadcuts, all you could do is drill a hole, or find natural streamcuts, which are few and far between."

"We as geologists are fortunate to live in a period of great road building."

"It's a way of sampling fresh rock. The road builders slice through indiscriminately, and no little rocks, no softer units are allowed to hide."

"A roadcut is to a geologist as a stethoscope is to a doctor."

"An X-ray to a dentist."

"The Rosetta Stone to an Egyptologist."

"A twenty-dollar bill to a hungry man."

"If I'm going to drive safely, I can't do geology."

In moist climates, where vegetation veils the earth, streamcuts are about the only natural places where geologists can see exposures of rock, and geologists have walked hundreds of thousands of miles in and beside streams. If roadcuts in the moist world are a kind of gift, they are equally so in other places. Rocks are not easy to read where natural outcrops are

Detail of road cut near Rock Springs, Wyoming.

Redwall limestone and small rapid on the Colorado River in Marble Canyon, Grand Canyon National Park, Arizona.

so deeply weathered that a hammer will virtually sink out of sight — for example, in piedmont Georgia. Make a fresh road-cut almost anywhere at all and geologists will close in swiftly, like missionaries racing anthropologists to a tribe just discovered up the Xingu.

Basin and Range

I USED TO SIT IN CLASS AND LISTEN TO the terms come floating down the room like paper airplanes. Geology was called a descriptive science, and with its pitted outwash plains and drowned rivers, its

hanging tributaries and starved coastlines, it was nothing if not descriptive. It was a fountain of metaphor — of isostatic adjustments and degraded channels, of angular unconformities and shifting divides, of rootless mountains and bitter lakes. Streams eroded headward, digging from two sides into mountain or hill, avidly struggling toward each other until the divide between them broke down, and the two rivers that did the breaking now became confluent (one yielding to the other, giving up its direction of flow and going the opposite way) to become a single stream. Stream capture. In the Sierra Nevada, the Yuba had captured the Bear. The Macho member of a formation in New Mexico was derived in large part from the solution and collapse of another formation. There was fatigued rock and incompetent rock and inequigranular fabric in rock. If you bent or folded rock, the inside of the curve was in a state of compression, the outside of the curve was under great tension, and somewhere in the middle was the surface of no strain. Thrust fault, reverse fault, normal fault — the two sides were active in every fault. The inclination of a slope on which boulders would stay put was the angle of repose. There seemed, indeed, to be more than a little of the humanities in this subject. Geologists communicated in English; and they could name things in a manner that sent shivers through the bones. They had roof pendants in their discordant batholiths, mosaic conglomerates in desert pavement. There was ultrabasic, deep-ocean, mottled green-and-black rock — or serpentine.

Salt patterns, Bonneville Salt Flats, Utah.

Playa near Wells, Nevada.

There was the slip face of the barchan dune. In 1841, a paleontologist had decided that the big creatures of the Mesozoic were "fearfully great lizards," and had therefore named them dinosaurs. There were festooned crossbeds and limestone sinks, pillow lavas and petrified trees, incised meanders and defeated streams. There were dike swarms and slickensides, explosion pits, volcanic bombs. Pulsating glaciers. Hogbacks. Radiolarian ooze. There was almost enough resonance in some terms to stir the

adolescent groin. The swelling up of mountains was described as an orogeny. Ontogeny, phylogeny, orogeny—accent syllable two. The Antler Orogeny, the Avalonian Orogeny, the Taconic, Acadian, Alleghenian Orogenies. The Laramide Orogeny. The center of the United States had had a dull geologic history— nothing much being accumulated, nothing much being eroded away. It was just sitting there conservatively. The East had once been radical—had been unstable, reformist, revolutionary, in the Paleozoic

pulses of three or four orogenies. Now, for the last hundred and fifty million years, the East had been stable and conservative. The far-out stuff was in the Far West of the country — wild, weirdsma, a leather-jacket geology in mirrored shades, with its welded tuffs and Franciscan mélange (internally deformed, complex beyond analysis), its strike-slip faults and falling buildings, its boiling springs and fresh volcanics, its extensional disassembling of the earth.

Basin and Range

Fossil taken from the Green River Formation, Wyoming, (now in the Ulrich Gallery near Fossil Butte National Monument).

Teton Range, Grand Teton National Park, Wyoming, reflected in the Snake River.

IN JUNE, EVERY YEAR, STUDENTS AND professors from Eastern colleges—with their hydrochloric-acid phials and their hammers and their Brunton compasses—head West. To be sure, there is plenty of absorbing geology under the shag of Eastern America, galvanic conundrums in Appalachian structure and intricate puzzles in history and stratigraphy. In no manner would one wish to mitigate the importance of the Eastern scene. Undeniably, though, the West is where the rocks are—"where it all hangs out," as someone in the United States Geological Survey has put it—and of Eastern geologists who do any kind of summer field work about seventy-five per cent go West. They carry state geological maps and the regional geological highway maps that have been published by the American Association of Petroleum Geologists—maps as prodigally colored as drip paintings and equally formless in their worm-trail-and-paramecium depictions of the country's

uppermost rock. The maps give two dimensions but more than suggest the third. They tell the general age and story of the banks of the asphalt stream. Karen Kleinspehn has been doing this for some years, getting into her Minibago, old and overloaded, a two-door Ford, heavy-duty springs, with odd pieces of the Rockies under the front seat and a mountain tent in the gear behind, to cross the Triassic lowlands and the Border Fault and to rise into the Ridge and Valley Province, the folded-and-faulted, deformed Appalachians—the beginnings of a journey that above all else is physiographic, a journey that tends to mock the idea of a nation, of a political state, as an unnatural subdivision of the globe, as a metaphor of the human ego sketched on paper and framed in straight lines and in riparian boundaries between unalterable coasts. The United States: really a quartering of a continent, a drawer in North America. Pull it out and prairie dogs would spill off one side, alli-

Ruin in Zion Canyon, Zion National Park, Utah.

gators off the other—a terrain crisscrossed with geological boundaries, mammalian boundaries, amphibian boundaries: the limits of the world of the river frog, the extent of the Nugget Formation, the range of the mountain cougar. The range of the cougar is the cougar's natural state, overlying segments of tens of thousands of other states, a few of them proclaimed a nation. The United States of America, with its capital city on the Atlantic Coastal Plain. The change is generally dramatic as one province gives way to another; and halfway across Pennsylvania, as you leave the quartzite ridges and carbonate valleys of the folded-and-faulted mountains, you drop for a moment into Cambrian rock near the base of a long climb, a ten-mile gradient upsection in time from the Cambrian into the Ordovician into the Silurian into the Devonian into the Mississippian (generally through the same chapters of the earth represented in the walls of the Grand Canyon) and finally out onto the Pennsylvanian itself, the upper deck, the capstone rock, of the Allegheny Plateau. Now even the Exxon map shows a new geology, roads running every which way like shatter lines in glass, following the crazed geometries of this deeply dissected country, whereas, before, the roads had no choice but to run northeast-southwest among the long ropy trends of the deformed mountains, following the endless ridges. On these transcontinental trips, Karen has driven as much as a thousand miles in a day at speeds that she has come to regard as dangerous and no less emphatically immoral. She has almost never slept under a roof, nor can she imagine why anyone on such a journey would want or need to; she "scopes out" her campsites in the late-failing light with strong affection for national forests and less for the three-dollar campgrounds where you roll out your Ensolite between two trailers, where gregarious trains honk like Buicks, and Yamahas on instruments climb escarpments in the night. The physiographic boundary is indistinct where you shade off the Allegheny Plateau and onto the stable craton, the continent's enduring core, its heartland, immemorially unstrained, the steady, predictable hedreocraton—the Stable Interior Craton. There are old mountains to the east, maturing mountains to the west, adolescent mountains beyond. The craton has participated on its edges in the violent creation of the mountains. But it remains intact within, and half a nation wide—the lasting, stolid craton, slowly, slowly downwasting. It has lost five centimetres since the birth of Christ. In much of Canada and parts of Minnesota and Wisconsin, the surface of the craton is Precambrian—earth-basement rock, the continental shield. Ohio, Indiana, Illinois, and so forth, the whole of what used to be called the Middle West, is shield rock covered with a sedimentary veneer that has never been metamorphosed, never been ground into tectonic hash—sandstones, siltstones, limestones, dolomites, flatter than the ground above them, the silent floors of departed oceans, of epicratonic seas. Iowa. Nebraska. Now with each westward township the country thickens, rises—a thousand, two thousand, five thousand feet—on crumbs shed off the Rockies and generously served to the craton. At last the Front Range comes to view—the chevroned mural of the mountains, sparkling white on gray, and on its outfanning sediments you are lifted into the Rockies and you plunge through a canyon to the Laramie Plains. "You go from one major geologic province to another and—whoa!—you really know you're doing it." There are mountains now

behind you, mountains before you, mountains that are set on top of mountains, a complex score of underthrust, upthrust, overthrust mountains, at the conclusion of which, through another canyon, you come into the Basin and Range. Brigham Young, when he came through a neighboring canyon and saw rivers flowing out on alluvial fans from the wall of the Wasatch to the flats beyond, made a quick decision and said, "This is the place." The scene suggested settling for it. The alternative was to press on beside a saline sea and then across salt barrens so vast and flat that when microwave relays would be set there they would not require towers. There are mountains, to be sure — off to one side and the other: the Oquirrhs, the Stansburys, the Promontories, the Silver Island Mountains. And with Nevada these high, discrete, austere new ranges begin to come in waves, range after range after north-south range, consistently in rhythm with

High Wyoming plains near South Pass City.

Great Salt Lake in Utah, seen from the Oquirrh Mountains.

wide flat valleys: basin, range; basin, range; a mile of height between basin and range. Beside the Humboldt you wind around the noses of the mountains, the Humboldt, framed in cottonwood—a sound, substantial, year-round-flowing river, among the largest in the world that fail to reach the sea. It sinks, it disappears, in an evaporite plain, near the bottom of a series of fault blocks that have broken out to form a kind of stairway that you climb to go out of the Basin and Range. On one step is Reno, and at the top is Donner Summit of the uplifting Sierra Nevada, which has gone above fourteen thousand feet but seems by no means to have finished its invasion of the sky. The Sierra is rising on its east side and is hinged on the west, so the slope is long to the Sacramento Valley—the physiographic province of the Great Valley—flat and sea-level and utterly incongruous within its

flanking mountains. It was not eroded out in the normal way of valleys. Mountains came up around it. Across the fertile flatland, beyond the avocados, stand the Coast Ranges, the ultimate province of the present, the berm of the ocean—the Coast Ranges, with their dry and straw-brown Spanish demeanor, their shadows of the live oaks on the ground.

Basin and Range

Wasatch quartzite, Twin Peaks Wilderness of the Wasatch Mountains, Utah.

DOMINY WAS BORN ON A FARM IN central Nebraska, and all through his youth his family and the families around them talked mainly of the vital weather. They lived close to the hundredth meridian, where, in a sense more fundamental than anything resulting from the events of United States history, the West begins. East of the hundredth meridian, there is enough rain to support agriculture, and west of it there generally is not. The Homestead Act of 1862, in all its promise, did not take into account this ineluctable fact. East of the hundredth meridian, homesteaders on their hundred and sixty acres of land were usually able to fulfill the dream that had been legislated for them. To the west, the odds against them were high. With local exceptions, there just was not enough water. The whole region between the hundredth meridian and the Rocky Mountains was at that time known as the Great American Desert. Still beyond the imagination were the ultramontane basins where almost no rain fell at all.

Growing up on a farm that had been homesteaded by his grandfather in the eighteen-seventies, Dominy often enough saw talent and energy going to waste under clear skies. The situation was marginal. In some years, more than twenty inches of rain would fall and harvests would be copious. In others, when the figure went below ten, the family lived with the lament that there was no money to buy clothes, or even sufficient food. These radical uncertainties were eventually removed by groundwater development, or reclamation—the storage of what

Eagle Falls in Eldorado National Forest of the Sierra Nevada.

water there was, for use in irrigation. When Dominy was eighteen years old, a big thing to do on a Sunday was to get into the Ford, which had a rumble seat, and go out and see the new dam. In his photo album he put pictures of reservoirs and irrigation projects. ("It was impressive to a dry-land farmer like me to see all that water going down a ditch toward a farm.") Eventually, he came to feel that there would be, in a sense, no West at all were it not for reclamation.

In Campbell County, Wyoming, the situation was not even marginal. This was high, dry country, suitable only for free-ranging livestock, not for farming. In the best of years, only about fourteen inches of rain might fall. "Streams ran water when the snow melted. Otherwise, the gulches were dry. It was the county with the most towns and the fewest people, the most rivers with the least water, and the most cows with the least milk in the world." It was, to the eye, a wide, expansive landscape with beguiling patterns of perspective. Its unending buttes, flat or nippled, were spaced out to the horizons like stone chessmen. Deer and antelope moved among them in herds, and on certain hilltops cairns marked the graves of men who had hunted buffalo. The herbage was so thin that forty acres of range could reasonably support only one grazing cow. Nonetheless, the territory had been homesteaded, and the homesteaders simply had not received from the federal government enough land for enough cattle to give them financial equilibrium as ranchers, or from the sky enough water to give them a chance as farmers. They were going backward three steps for each two forward. Then the drought came.

"Nature is a pretty cruel animal. I watched the people there—I mean good folk, industrious, hardworking, frugal—

compete with the rigors of nature against hopeless odds. They would ruin their health and still fail." Without waiting for approval from Cheyenne or Washington, the young county agent took it upon himself to overcome nature if the farmers and ranchers could not. He began up near Recluse, on the ranch of a family named Oedekoven, in a small bowl of land where an intermittent stream occasionally flowed. With a four-horse Fresno —an ancestral bulldozer—he moved earth and plugged the crease in the terrain where the water would ordinarily run out and disappear into the ground and the air. He built his little plug in the classic form of the earth-fill dam—a three-for-one slope on the water side and two-for-one the other way. More cattle died, but a pond slowly filled, storing water. The pond is still there, and so is Oedekoven, the rancher.

For two and a half years, Dominy lived with his wife and infant daughter in a stone dugout about three miles outside Gillette, the county seat. For light they used a gasoline lantern. For heat and cooking they had a coal-burning stove. Dominy dug the coal himself out of a hillside. His wife washed clothes on a board. On winter mornings when the temperature was around forty below zero, he made a torch with a rag and a stick, soaked it in kerosene, lighted it, and put it under his car. When the car was warm enough to move, Dominy went off to tell ranchers and farmers about the Corn-Hog Program ("Henry Wallace slaughtering piglets to raise the price of ham"), the Wheat Program (acreage control), or how to build a dam. "Campbell County was my kingdom. When I was twenty-four years old, I was king of the god-damned county." He visited Soda Well, Wild Cat, Teckla, Turnercrest—single-family post offices

widely spaced—or he followed the farmers and ranchers into the county seat of the county seat, Jew Jake's Saloon, where there was a poker game that never stopped and where the heads of moose, deer, elk, antelope, and bighorn sheep looked down on him and his subjects, feet on the rail at 9 A.M. Dominy had his first legitimate drink there. The old brass rail is gone—and so is Dominy—but the saloon looks just the same now, and the boys are still there at 9 A.M.

There was an orange scoria butte behind Dominy's place and an alfalfa field in front of it. Rattlesnakes by the clan came out of the butte in the spring, slithered around Dominy's house, and moved on into the alfalfa for the summer. In September, the snakes headed back toward the butte. Tomatoes were ripe in Dominy's garden, and whenever he picked some he first took a hoe and cleared out the rattlesnakes under the vines. Ranchers got up at four in the morning, and sometimes Dominy was outside honking his horn to wake them. He wanted them to come out and build dams—dams, dams, dams. "I had the whole county stirred up. We were moving! Stockpond dam and reservoir sites were supposed to be inspected first by Forest Service rangers, but who knows when they would have come? I took it upon myself to ignore these pettifogging minutiae." Changing the face of the range, he polka-dotted it with ponds. Dominy and the ranchers and farmers built a thousand dams in one year, and when they were finished there wasn't a thirsty cow from Jew Jake's Saloon to the Montana border. "Christ, we did more in that county in one year than any other county in the country. That range program really put me on the national scene."

Encounters with the Archdruid

Pond along the
Green River, Wind
River Range,
Wyoming.

2. HOMESTEADING

LAND WAS A FORM OF RELIGION to the Indians, and the Black Hills, in this sense, were the religion of the Sioux. With all the fish, game, and beauty any man could want, the Black Hills fed the Sioux in body and spirit. Indians had no sense of private property, private land. The idea of individual human beings' owning pieces of the earth was to them at first incomprehensible and, when comprehended, a form of sacrilege. With the white man and his sense of property and the rights of property came the inequities and paradoxes that eventually led to the need for a conservation movement. Meanwhile, in 1851 the Sioux were promised by treaty that they could keep their Black Hills forever. In 1874, white men found gold there, and in 1875 white men entered the Black Hills in staggering numbers—white trash, in the main, like Wild Bill Hickok. It was the last gold rush in the United States. The promise to the Sioux was permanently broken, and the Sioux expressed their grief by destroying General Custer and his soldiers. "The Sioux are now a hundred miles east of here on a flat reservation in the Badlands," Park said to me. "There are no Sioux in the Sierra Club."

Encounters with the Archdruid

ON THE WAY UP TO THE LOOKOFF, WE had stopped at a spring, where I buried my face in watercress and simultaneously drank and ate. Love said that F. V. Hayden, the first reconnaissance geologist in Wyoming Territory, also happened to be a medical doctor, and he went around dropping watercress in springs and streams to prevent scurvy from becoming the manifest destiny of emigrants. Hayden, who taught at the University of Pennsylvania, led one of the several groups that in 1879 combined to become the United States Geological Survey. When he came into the country in the late eighteen-fifties, he was so galvanized by seeing the composition of the earth in clear unvegetated view that he regularly went off on his own, moved hurriedly from outcrop to outcrop, and filled canvas bags with samples. This puzzled the Sioux. Wondering what he could be collecting, they watched him, discussed him, and finally attacked him. Seizing his canvas bags, they shook out the contents. Rocks fell on the ground. In that instant, Professor Hayden was accorded the special sta-

Aerial view of Badlands, Wyoming.

Badlands on the Wyoming plains.

tus that all benevolent people reserve for the mentally disadvantaged. In their own words, the Sioux named him He Who Picks Up Rocks Running, and to all hostilities thereafter Hayden remained immune.

Rising from the Plains

IN THE CENTER OF THE HIGH SWALE were the silvery-gray remains of hundreds of cut trees, which had been dragged into the open and arranged as a fence in kidney-bean shape, all but enclosing about fifteen acres. Vaguely, they formed a double corral, with an aperture in one place only, and had apparently been used — for uncounted years — to trap antelopes. Antelopes don't climb fences, as people fond of roast pronghorn discovered centuries ago. Love's son Charlie, the professor of anthropology and geology at Western Wyoming Community College, knew of

the trap and had thought out the strategies by which it was effective. His father expressed pride in Charlie for "thinking as intelligently as the aborigines." The high valley held fast an aesthetic silence—a silence reminiscent of the Basin and Range, a silence equal to the winter Yukon. About the only sign of humankind was the antelope trap. This was the Overthrust Belt as it had appeared before white people—thinking intelligently but not like the aborigines—

mapped the terrain, modelled its structure, and went after what lay beneath it. There were mountain bluebells and salt sage in the valley, ground phlox and prickly pear. Love reached down and plucked up a plant and asked if I knew what it was. It looked familiar, and I said, "Wild onion."

He said, "It's death camas. It brings death quickly. It killed many pioneer children. They thought it looked like wild onion."

Rising from the Plains

Prickly pear cactus.

Upper Geyser Basin, Yellowstone National Park, Wyoming.

IN THE FALL OF 1865, MAJOR GENERAL Grenville Dodge and his pack trains and cavalry and other troops were coming south along the St. Vrain Trail, under the front of the Laramie Range. The Powder River campaign, behind them, had been, if not a military defeat, a signal failure in its purpose: to cow the North Cheyennes and the Ogallala Sioux. General Dodge, though, was preoccupied with something else. President Lincoln, not long before he died, had instructed Dodge to choose a route for the Union Pacific Railroad. Dodge, like others before him, had sought the counsel of Jim Bridger, the much celebrated trapper, explorer, fur trader, commercial entrepreneur, and all-around mountain man. Bridger, who was sixty by then, had preceded almost everybody else into the West by two or three decades and knew the country as few other whites ever would. It was he who discovered the Great Salt Lake, reporting his find as the Pacific Ocean. It was he whose descriptions of Jackson Hole, Yellowstone Lake, Yellowstone Falls, the Fire Hole geysers, and the Madison River had once been known as "Jim Bridger's lies." His father-in-law was Chief Washakie. And now this bluecoat general wanted to know where to put a railroad. The Oregon Trail went around the north end of the Laramie

Firehole River
Falls, Yellowstone
National Park.

Range and up the Sweetwater to South Pass—to say the least, an easy grade. But for a competitive transcontinental railroad the Sweetwater was a route of wide digression and no coal. Bridger mentioned Lodgepole Creek and said the high ground above it was the low point on the crest of the Laramie Range (a fact that theodolites would in time confirm). The route could go there.

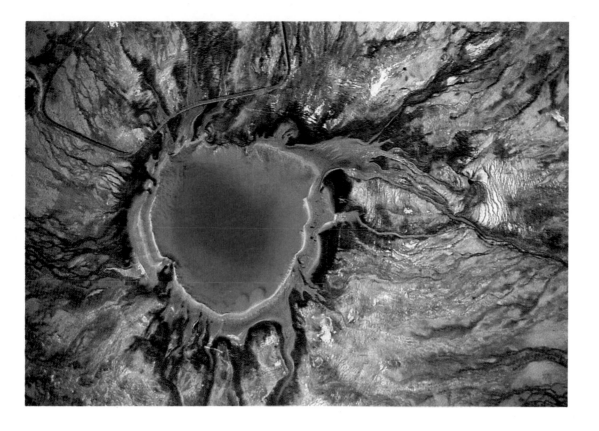

Yellowstone Lake.

So Dodge, in 1865, coming south from the Powder River, left his pack trains and cavalry on the St. Vrain Trail and led a small patrol up Lodgepole Creek. At the top, he turned south and did reconnaissance of the summit terrain. In the small valley of a high tributary of Crow Creek—five or ten miles south of Bridger's recommendation—he surprised a band of Indians. His report does not say of what tribe. They were hostiles—or at least became so after Dodge started firing at them. At the moment of mutual surprise, they were between him and his main column, and that made him tactically nervous. The patrol dismounted and walked due east—"holding the Indians at bay, when they came too near, with our Winchesters." In this manner, the gangplank was discovered. As Dodge kept going east, expecting to reach the escarpment from which he would signal with smoke, he reached no escarpment. Instead, he reached the remnant of the high ancient surface—this interfluvial isthmus between Crow Creek and Lone Tree Creek—touching the mountain summit.

It led down to the plains without a break. I then said to my guide that if we saved our scalps I believed we had found the crossing.

General Dodge went back east, and in the spring of 1867 returned with his route approved. The Union Pacific at that time ended in the middle of Nebraska. He got off the train, went up the North Platte, up the Lodgepole, and, as he approached the mountains, went directly overland to Crow Creek, where he staked out the western end of the railroad's next division. Without much pleasing anybody, he named the place Cheyenne. In no time, he was defending himself against furious Cheyennes. They killed soldiers and laborers, pulled up survey stakes, stole animals, and destroyed equipment. When some politicians, bureaucrats, and financiers arrived on a see-it-yourself junket west, the Cheyennes attacked them. With drawn revolver, General Dodge told his visitors, "We've got to clean these damn Indians out or give up building the Union Pacific Railroad. The government may take its choice."

The narrowest point on the gangplank is wide enough for the Union Pacific and nothing else. The interstate highway clings to one side. The tracks and lanes are so close that the gangplank resembles the neck of a guitar. A long coal freight slid by us. "The coal isn't piled higher than the tops of the gondolas," Love commented. "It's an environmental move—to keep the dust from blowing downwind." He said it was a good idea, no doubt, but he had experienced so many cinder showers earlier in his life that he could not help thinking that this latter-day assault on dust was "like bringing a fire under control at timberline." A cinder shower was what happened when an old-time locomotive pulled into a town and blew its stack. He also said that this could not have been an important emigrant route, because there was a lack of grass and water—absolute necessities for animal-powered travel. To the Union Pacific, however, the gangplank offered speed, efficiency, and hence predominance with respect to the competition. When the Denver & Rio Grande was laboring up switchbacks in a hamper-

ing expenditure of money and time—and the Santa Fe was struggling not only with mountains but also with desert terrain—the Union Pacific had already run up the gangplank, opened the West, and become everybody's Uncle Pete. Love said, "Out here, Uncle Sam is a gnat under a blanket compared to Uncle Pete. The Union Pacific had the best of it. This Miocene Ogallala formation was the youngest of the high-plains deposits that lapped onto the mountain front. It's subtle and seems academic until you try to build a railroad. This is the only place in the whole Rocky Mountain front where you can go from the Great Plains to the summit of the mountains without snaking your way up a mountain face or going through a tunnel. This one feature had more to do with the building of the West than any other factor. I don't diminish the importance of the Oregon Trail, but here you had everything going for you. This point hasn't been made before."

When the railroad was built, it was given (by the federal government) fifty per cent of the land in a forty-mile swath along its route—in checkerboard fashion, one square mile in every two. Today, Uncle Pete owns, among many things, the Rocky Mountain Energy Company, the Upland Industries Corporation, the Champlin Petroleum Company, and enough unmined uranium to send Wyoming to the moon. In Cheyenne, the Union Pacific station and the state capitol face each other at opposite ends of Capitol Avenue. The Union Pacific station came out of the Laramie Range, forty miles west, and, like the range itself, is sheathed in the russet Pennsylvanian sandstone and has a foundation of Precambrian granite. At least as imposing as the capitol, it is a baronially escutcheoned mountain of grandeur.

Indians, of course, had used the gangplank for who knows how long before General Dodge surprised them on the Laramie summit. They had crossed it on their journeys from the Great Plains to the Laramie Basin and on up to hunting grounds in the Medicine Bow Mountains. And the Indians, from the beginning, were themselves following a trail. Buffalo discovered the gangplank. "It was a buffalo trail," Love said. "Buffalo were the real trailmakers—trails you wouldn't believe. They were as good as the best civil engineers. It remains true today. If you're in Yellowstone, in the backcountry, and you have trouble finding your way across swamps, mountains, and thermal areas, you look for a buffalo trail and you'll get through." Beside Interstate 80 on the gangplank, a sign said:

"GAME CROSSING."

Rising from the Plains

THE *SHOSHONE PATHFINDER,* IN Lander, published in 1906 a special issue urging young people to make their lives in central Wyoming. "We beg leave to extend to each and every one of you a most cordial welcome to come, remain, and help develop a country so rich in natural resources as to be beyond the computation of mortal man," wrote the publishers. It was a country "clothed in a mantle of the most nutritious grasses and sage brush browse." In its Wind River Mountains were "thousands of square miles of dense forests, which the foot of man has never invaded, and . . . as to the supply and quality of timber in this county it will meet the requirements of all demands for all time to come." Moreover, there was coal: "It has been said of our coal fields that the entire United States would be unable to exhaust them in a century

. . . . It is in excess of the imagination to contemplate the vastness of this tremendous supply of fuel or what would ever transpire to exhaust it." And there was oil: "It is a recognized fact of long standing that the quantity of oil stored in the natural reservoirs of this county is so great that no estimate can be made." And there was gold. At the south end of the Wind Rivers, nearly five million dollars had come out of small mines with names like Hard Scrabble, Ground Hog, Hidden Hand, Mormon Crevice, Iron Duke, Midget, Rustler, Cariboo, and Irish Jew. "None of the mines have been exhausted, but merely sunk to a depth where more and better machinery is required." There was uranium, too, but as yet no compelling need to find it, and as yet no geologist equal to the task.

Rising from the Plains

MISS WAXHAM'S SCHOOL WAS A LOG cabin on Twin Creek near the mouth of Skull Gulch, a mile from the Mills ranch. Students came from much greater distances, even through deep snow. Many mornings, ink was frozen in the inkwells, and the day began with ink-thawing, followed by reading, spelling, chemistry, and civil government. Sometimes snow blew through the walls, forming drifts in the schoolroom. Water was carried from the creek—drawn from a hole that was chopped in the ice. If the creek was frozen to the bottom, the students melted snow. Their school was fourteen by sixteen feet—smaller than a bathroom at Wellesley. The door was perforated with bullet holes from "some passerby's six-shooter." Over the ceiling poles were old gunnysacks and overalls, to prevent the sod roof from shedding sediment on the students. Often, however, the air sparkled

Pioneer wagon and cabins, South Pass City, Wyoming.

with descending dust, struck by sunlight coming in through the windows, which were all in the south wall. There was a table and chair for Miss Waxham, and eight desks for her pupils. Miss Waxham's job was to deliver a hundred per cent of the formal education available in District Eleven, Fremont County, Wyoming.

The first fifteen minutes or half hour are given to reading "Uncle Tom's Cabin" or "Kidnapped," while we all sit about the stove to keep warm. Usually in the middle of a reading the sound of a horse galloping down the frozen road distracts the attention of the boys, until a few moments later six-foot George opens the door, a sack of oats in one hand, his lunch tied up in a dish rag in the other. Cold from his five-mile ride, he sits down on the floor by the stove, unbuckles his spurs, pulls off his leather chaps, drops his hat, unwinds two or three red handkerchiefs from about his neck and ears, takes off one or two coats, according to the temperature, unbuttons his vest and straightens his leather cuffs. At last he is ready for business.

Sandford is the largest scholar, six feet, big, slow in the school room, careful of every move of his big hands and feet. His voice is subdued and full of awe as he calls me "ma'am." Outside while we play chickens he is another person—there is room for his bigness. Next largest of the boys is Otto Schlicting, thin and dark, a strange combination of shrewdness and stupidity. His problems always prove, whether they are right or not! He is a boaster, too, tries to make a big impression. But there is something very attractive about him. I was showing his little sister how to add and subtract by making little lines and adding or crossing off others. Later I found on the back of Otto's papers hundreds and hundreds of little lines—trying to add that way as far as a hundred evidently. He is nearly fifteen and studying division. . . . Arithmetic is the family failing. "How many eights in ninety-six?" I ask him. He thinks for a long time. Finally he says—with such a winsome smile that I wish with all my heart it were true— "Two." "What feeds the cells in your body?" I ask him. He thinks. He says, "I guess it's vinegar." He has no idea of form. His maps of North America on the board are all like turnips.

Students' ages ranged through one and two digits, and their intelligence even more widely. When Miss Waxham called upon Emmons Schlicting, asking, "Where does digestion take place?," Emmons answered, "In the Erie Canal." She developed a special interest in George Ehler, whose life at home was troubled.

He is only thirteen, but taller than Sandford, and fair and handsome. I should like to get him away from his family—kidnap him. To think that it was he who tried to kill his father! His face is good as can be.

At lunchtime, over beans, everyone traded the news of the country, news of whatever might have stirred in seven thousand square miles: a buffalo wolf trapped by Old Hanley; missing horses and cattle, brand by brand; the sheepherder most recently lost in a storm. If you went up Skull Gulch, behind the school, and climbed to the high ground beyond, you could see seventy, eighty, a hundred miles. You "could see the faint outlines of Crowheart Butte, against the Wind River Range." There was a Wyoming-history lesson in the naming of Crowheart Butte, which rises a thousand feet above the surrounding landscape and is capped with flat sandstone. To this day, there are tepee rings on Crowheart Butte. One of the more arresting sights in remote parts of the West are rings of stones that once resisted the wind and now recall what blew away. The Crows liked the hunting country in the area of the butte, and so did the Shoshonis. The two tribes fought, and lost a lot of blood, over this ground. Eventually, the chief of the Shoshonis said, in effect, to the chief of the Crows: this is pointless; I will fight you, one against one; the hunting ground goes to the winner. The chief of the Shoshonis was the great Washakie, whose name rests in six places on the map of Wyoming, including a mountain range and a county. Washakie was at least fifty, but fit. The Crow would have been wise to demur. Washakie destroyed him in the hand-to-hand combat, then cut out his heart and ate it.

Despite her relative disadvantages as a newcomer, an outlander, and an educational ingénue, Miss Waxham was a quick study. Insight was her long suit, and in no time she understood Wyoming. For example, an entry in her journal says of George Ehler's father, "He came to the country with one mare. The first summer, she had six colts! She must have had calves, too, by the way the Ehlers' cattle increased." These remarks were dated October 22, 1905—the day after her stagecoach arrived. In months that followed, she sketched her neighbors (the word applied over many tens of miles). "By the door was Mrs. Frink, about 18, with Frink junior, a large husky baby. Ida Franklin, Mrs. Frink's sister and almost her double, was beside her, frivolous even in her silence." There was the story of Dirty Bill Collins, who had died as a result of taking a bath. And she fondly recorded Mrs. Mills' description of the libertine Guy Signor: "He has a cabbage heart with a leaf for every girl." She noted that the nearest barber had learned his trade shearing sheep, and a blacksmith doubled as dentist. Old Pelon, a French Canadian, impressed her, because he had refused to ask for money from the government after Indians killed his brother. "Him better dead," said Old Pelon. Old Pelon was fond of the masculine objective pronoun. Miss Waxham wrote, "Pelon used to have a wife, whom he spoke of always as 'him.'" Miss Waxham herself became a character in this tableau. People sometimes called her the White-Haired Kid.

"There's many a person I should be glad to meet," read an early entry in her journal. She wanted to meet Indian Dick, who had been raised by Indians and had no idea who he was—probably the orphan of emigrants the Indians killed. She wanted to meet "the woman called Sour Dough; Three Fingered Bill, or Suffering Jim; Sam Omera, Reub Roe. . . ." (Reub Roe held up wagons and stagecoaches looking for members of the Royal Family.) Meanwhile, there was one flockmaster and itinerant cowboy who seemed more than pleased to meet her.

In the first reference to him in her journal she calls him "Mr. Love—Johnny Love." His place was sixty miles away, and he had a good many sheep and cattle to look after, but somehow he managed to be right there when the new young schoolmarm arrived. In the days, weeks, and months that followed, he showed a pronounced tendency to reappear. He came, generally, in the dead of night, unexpected. Quietly he slipped into the corral, fed and watered his horse, slept in the bunkhouse, and was there at the table for breakfast in the morning—this dark-haired, blue-eyed, handsome man with a woolly Midlothian accent.

Rising from the Plains

IN THE UNITED STATES GEOLOGICAL Survey's seven-and-a-half-minute series of topographic maps is a quadrangle named Love Ranch. The landscape it depicts lies just under the forty-third parallel and west of the hundred-and-seventh meridian—coordinates that place it twelve miles from the geographic center of Wyoming. The names of its natural features are names that more or less materialized around the kitchen table when David Love was young: Corral Draw, Castle Gardens, Buffalo Wallows, Jumping-Off Draw. To the fact that he grew up there his vernacular, his outlook, his pragmatic skills, and his professional absorptions about equally attest. The term "store-bought" once brightened his eyes. When

Pioneer artifacts at Rock Creek Ranch, Desolation Canyon, Utah.

one or another of the cowpunchers used a revolver, the man did not so much fire a shot as "slam a bullet." If a ranch hand was tough enough, he would "ride anything with hair on it." Coffee had been brewed properly if it would "float a horseshoe." Blankets were "sougans." A tarpaulin was a "henskin." To be off in the distant ranges was to be "gouging around the mountains." In Love's stories of the ranch, horses come and go by the "cavvy." If they are unowned and untamed, they are a "wild bunch"—led to capture by a rider "riding point." In the flavor of his speech the word "ornery" endures.

He describes his father as a "rough, kindly, strong-willed man" who would put a small son on each knee and—reciting "Ride a cockhorse to Banbury Cross to see a fine lady upon a white horse"—give the children bronco rides after dinner, explaining that his purpose was "to settle their stomachs." Their mother's complaints went straight up the stovepipe and away with the wind. When their father was not reciting such Sassenach doggerel, he could draw Scottish poems out of the air like bolts of silk. He had the right voice, the Midlothian timbre. He knew every syllable of "The Lady of the Lake." Putting his arms around the shoulders of his wee lads, he would roll it to them by the canto, and when they tired of Scott there were in his memory more than enough ballads to sketch the whole of Scotland, from the Caithness headlands to the Lammermuir Hills.

David was fifteen months younger than his brother, Allan. Their sister, Phoebe, was born so many years later that she does not figure in most of these scenes. They were the only children in a thousand square miles, where children outnumbered the indigenous trees. From the ranch buildings, by Muskrat Creek, the Wind River Basin reached out in buffalo grass, grama grass, and edible salt sage across the cambered erosional swells of the vast dry range. When the wind dropped, this whole wide world was silent, and they could hear from a great distance the squeak of a horned lark. The nearest neighbor was thirteen miles away. On the clearest night, they saw no light but their own.

Old buffalo trails followed the creek and branched from the creek: old but not ancient—there were buffalo skulls beside them, and some were attached to hide. The boys used the buffalo trails when they rode off on ranch chores for their father. They rode young and rode long, and often went without water. Even now, six decades later, David will pass up a cool spring, saying, "If I drink now, I'll be thirsty all day." To cut cedar fence posts, they went with a wagon to Green Mountain, near Crooks Gap—a round trip of two weeks. In early fall, each year, they spent ten days going back and forth to the Rattlesnake Hills for stove wood. They took two wagons—four horses pulling each wagon—and they filled them with limber pine. They used axes, a two-handled saw. Near home, they mined coal with their father—from the erosional wonderland they called Castle Gardens, where a horse-drawn scraper stripped the overburden and exposed the seams of coal. Their father was adept at corralling wild horses, a skill that called for a horse and rider who could outrun these closest rivals to the wind. He caught more than he kept, put his Flatiron brand on the best ones and sold the others. Some of them escaped. David remembers seeing one clear a seven-foot bar in the wild-horse corral and not so much as touch it. When he and Allan were in their early teens, his father sent them repping—representing Love

Ranch in the general roundup — and they stayed in cow camp with other cowboys, and often enough their sougans included snow. When they were out on the range, they slept out on the range, never a night in a tent. This was not a choice. It was a family custom.

In the earlier stretch of his life when John Love had slept out for seven years, he would wrap himself in his sougans and finish the package with the spring hooks and D-rings that closed his henskin. During big gales and exceptional blizzards, he looked around for a dry wash and the crease of an overhanging cutbank. He gathered sage and built a long fire — a campfire with the dimensions of a cot. He cooked his beans and bacon, his mutton, his sourdough, his whatever. After dinner, he kicked the fire aside and spread out his bedroll. He opened his waterproof packet of books and read by kerosene lamp. Then he blew out the light and went to sleep on warm sand. His annual expenditures were seventy-five dollars. This was a man who wore a long bear-

Snow drifts in Central Wyoming plains.

skin coat fastened with bone pegs in loops of rope. This was a man who, oddly enough, carried with him on the range a huge black umbrella—his summer parasol. This was a man whose Uncle John Muir had invented a device that started a fire in the morning while the great outdoorsman stayed in bed. And now this wee bairn with the light-gold hair was, in effect, questioning Love Ranch policy by asking his father what he had against tents. "Laddie, you don't always have one available," his father said patiently. "You want to get used to living without it." Tents, he made clear, were for a class of people he referred to as "pilgrims."

When David was nine, he set up a trap line between the Hay Meadow and the Pinnacles (small sandstone buttes in Castle Gardens). He trapped coyotes, bobcats, badgers. He shot rabbits. He ran the line on foot, through late-autumn and early-winter snow. His father was with him one cold and blizzarding January day when David's rifle and the rabbits he was carrying slipped from his hands and fell to the snow. David picked up the gun and soon dropped it again. "It was a cardinal sin to drop a rifle," he says. "Snow and ice in the gun barrel could cause the gun to blow up when it was fired." Like holding on to a saddle horn, it was something you just did not do. It would not have crossed his father's mind that David was being careless. In sharp tones, his father said, "Laddie, leave the rabbits and rifle and run for home. Run!" He knew hypothermia when he saw it, no matter that it lacked a name.

Even in October, a blizzard could cover the house and make a tunnel of the front veranda. As winter progressed, rime grew on the nailheads of interior walls until white spikes projected some inches into the rooms. There were eleven rooms.

His mother could tell the outside temperature by the movement of the frost. It climbed the nails about an inch for each degree below zero. Sometimes there was frost on nailheads fifty-five inches up the walls. The house was chinked with slaked lime, wood shavings, and cow manure. In the wild wind, snow came through the slightest crack, and the nickel disks on the dampers of the heat stove were constantly jingling. There came a sound of hooves in cold dry snow, of heavy bodies slamming against the walls, seeking heat. John Love insulated his boots with newspapers—as like as not the *New York Times*. To warm the boys in their beds on cold nights, their mother wrapped heated flatirons in copies of the *New York Times*. The family were subscribers. Sundays only. The *Times*, David Love recalls, was "precious." They used it to insulate the house: pasted it against the walls beside the *Des Moines Register,* the *Tacoma News Tribune*—any paper from anywhere, without fine distinction. With the same indiscriminate voracity, any paper from anywhere was first read and reread by every literate eye in every cow camp and sheep camp within tens of miles, read to shreds and passed along, in tattered circulation on the range. There was, as Love expresses it, "a starvation of print." Almost anybody's first question on encountering a neighbor was "Have you got any newspapers?"

The ranch steadings were more than a dozen buildings facing south, and most of them were secondhand. When a stage route that ran through the ranch was abandoned, in 1905, John Love went down the line shopping for moribund towns. He bought Old Muskrat—including the hotel, the post office, Joe Lacey's Muskrat Saloon—and moved the whole of it eighteen miles. He bought

Golden Lake and moved it thirty-three. He arranged the buildings in a rough semicircle that embraced a corral so large and solidly constructed that other ranchers travelled long distances to use it. Joe Lacey's place became the hay house, the hotel became in part a saddlery and cookhouse, and the other buildings, many of them connected, became all or parts of the blacksmith shop, the chicken hatchery, the ice shed, the buggy shed, the sod cellar, and the bunkhouse—social center for all the workingmen from a great many miles around. There was a granary made of gigantic cottonwood logs from the banks of the Wind River, thirty miles away. There were wool-sack towers, and a wooden windmill over a hand-dug well. The big house itself was a widespread log collage of old town parts and original construction. It had wings attached to wings. In the windows were air bubbles in distorted glass. For its twenty tiers of logs, John had journeyed a hundred miles to the lodgepole-pine groves of the Wind River Range, returning with ten logs at a time, each round trip requiring two weeks. He collected a hundred and fifty logs. There were no toilets, of course, and the family had to walk a hundred feet on a sometimes gumbo-slick path to a four-hole structure built by a ranch hand, with decorative panelling that matched the bookcases in the house. The cabinetmaker was Peggy Dougherty, the stagecoach driver who had first brought Miss Waxham through Crooks Gap and into the Wind River country.

The family grew weary of carrying water into the house from the well under the windmill. And so, as she would write in later years:

After experiments using an earth auger and sand point, John triumphantly installed a pitcher pump in the kitchen, a sink, and drain pipe to a barrel, buried in the ground at some distance from the house. This was the best, the first, and at that time the only water system in an area the size of Rhode Island.

In the evenings, kerosene lamps threw subdued yellow light. Framed needlework on a wall said "WASH & BE CLEAN." Everyone bathed in the portable galvanized tub, children last. The more expensive galvanized tubs of that era had built-in seats, but the Loves could not afford the top of the line. On the plank floor were horsehide rugs—a gray, a pinto—and the pelt of a large wolf, and two soft bobcat rugs. Chairs were woven with rawhide or cane. John recorded the boys' height on a board nailed to the inside of the kitchen doorframe. A brass knocker on the front door was a replica of a gargoyle at Notre-Dame de Paris.

The family's main sitting and dining room was a restaurant from Old Muskrat. On the walls were polished buffalo horns mounted on shields. The central piece of furniture was a gambling table from Joe Lacey's Muskrat Saloon. It was a poker-and-roulette table—round, covered with felt. Still intact were the subtle flanges that had caused the roulette wheel to stop just where the operator wished it to. And if you reached in under the table in the right place you could feel the brass slots where the dealer kept wild cards that he could call upon when the fiscal integrity of the house was threatened. If you put your nose down on the felt, you could almost smell the gunsmoke. At this table David Love received his basic education—his schoolroom a restaurant, his desk a gaming table from a saloon. His mother may have been trying to academize the table when she covered it with a red-and-white India print.

From time to time, other school-marms were provided by the district. They came for three months in summer. One came for the better part of a year. By and large, though, the boys were taught by their mother. She had a rolltop desk, and Peggy Dougherty's glassed-in book-cases. She had the 1911 Encyclopedia Britannica, the Redpath Library, a hundred volumes of Greek and Roman literature, Shakespeare, Dickens, Emerson, Thoreau, Longfellow, Kipling, Twain. She taught her sons French, Latin, and a bit of Greek. She read to them from books in German, translating as she went along. They read the Iliad and the Odyssey. The room was at the west end of the ranch house and was brightly illuminated by the setting sun. When David as a child saw sunbeams leaping off the books, he thought the contents were escaping.

In some ways, there was more chaos in this remote academic setting than there could ever be in a grade school in the heart of a city.

The house might be full of men, waiting out a storm, or riding on a round-up. I was baking, canning, washing clothes, making soap. Allan and David stood by the gasoline washing machine reading history or geography while I put sheets through the wringer. I ironed. They did spelling beside the ironing board, or while I kneaded bread; they gave the tables up to 15 times 15 to the treadle of the sewing machine. Mental problems, printed in figures on large cards, they solved while they raced across the . . . room to write the answers . . . and learned to think on their feet. Nine written problems done correctly, without help, meant no tenth problem. . . . It was surprising in how little time they finished

their work — to watch the butchering, to help drive the bawling calves into the weaning pen, or to get to the corral, when they heard the hoofbeats of running horses and the cries of cowboys crossing the creek.

No amount of intellectual curiosity or academic discipline was ever going to hold a boy's attention if someone came in saying that the milk cow was mired in a bog hole or that old George was out by the wild-horse corral with the biggest coyote ever killed in the region, or if the door opened and, as David recalls an all too typical event, "they were carrying in a cowboy with guts ripped out by a saddle horn." The lessons stopped, the treadle stopped, and she sewed up the cowboy.

Across a short span of time, she had come a long way with these bunkhouse buckaroos. In her early years on the ranch, she had a lesser sense of fitting in than she would have had had she been a mare, a cow, or a ewe. She did not see another woman for as much as six months at a stretch, and if she happened to approach a group of working ranch hands they would loudly call out, "Church time!" She found "the sudden silence . . . appalling." Women were so rare in the country that when she lost a glove on the open range, at least twenty miles from home, a stranger who found it learned easily whose it must be and rode to the ranch to return it. Men did the housekeeping and the cooking, and went off to buy provisions at distant markets. Meals prepared in the bunkhouse were carried to a sheep wagon, where she and John lived while the big house was being built and otherwise assembled. The Wyoming sheep wagon was the ancestral Winnebago. It had a spring bed and a kitchenette.

After her two sons were born and became old enough to coin phrases, they

called her Dainty Dish and sometimes Hooty the Owl. They renamed their food, calling it, for example, dog. They called other entrées caterpillar and coyote. The kitchen stool was Sam. They named a Christmas-tree ornament Hopping John. It had a talent for remaining unbroken. They assured each other that the cotton on the branches would not melt. David decided that he was a camel, but later changed his mind and insisted that he was "Mr. and Mrs. Booth." His mother described him as "a light-footed little elf."

She noted his developing sense of scale when he said to her, "A coyote is the whole world to a flea."

One day, he asked her, "How long does a germ live?"

She answered, "A germ may become a grandfather in twenty minutes."

He said, "That's a long time to a germ, isn't it?"

She also made note that while David was the youngest person on the ranch he was nonetheless the most adroit at spotting arrowheads and chippings.

Sheepherder's camp near Fontenelle, Wyoming.

When David was five or six we began hunting arrowheads and chippings. While the rest of us labored along scanning gulches and anthills, David rushed by chattering and picking up arrowheads right and left. He told me once, "There's a god of chippings that sends us anthills. He lives in the sky and tinkers with the clouds."

The cowboys competed with Homer in the entertainment of Allan and David. There was one who—as David remembers him—"could do magic tricks with a lariat rope, making it come alive all around his horse, over our heads, under our feet, zipping it back and forth around us as we jumped up and down and squealed with delight." Sombre tableaux, such as butcherings, were played out before them as well. Years later, David would write in a letter:

We always watched the killing with horror and curiosity, although we were never permitted to participate at that age. It seemed so sad and so irrevocable to see the gushing blood when throats were cut, the desperate gasps for breath through severed windpipes, the struggle for and the rapid ebbing of life, the dimming and glazing of wide terrified eyes. We realized and accepted the fact that this was one of the procedures that were a part of our life on the range and that other lives had to be sacrificed to feed us. Throat-cutting, however, became a symbol of immediate death in our young minds, the ultimate horror, so dreadful that we tried not to use the word "throat."

He has written a recollection of the cowboys, no less frank in its bequested fact, and quite evidently the work of the son of his mother.

The cowboys and horse runners who drifted in to the ranch in ever-increasing numbers as the spring advanced were lean, very strong, hard-muscled, taciturn bachelors, nearly all in their twenties and early thirties. They had been born poor, had only rudimentary education, and accepted their lot without resentment. They worked days that knew no hour limitations but only daylight and dark, and weeks that had no holidays. . . . Most were homely, with prematurely lined faces but with lively eyes that missed little. None wore glasses; people with glasses went into other kinds of work. Many were already stooped from chronic saddle-weariness, bowlegged, hip-sprung, with unrepaired hernias that required trusses, and spinal injuries that required a "hanging pole" in the bunkhouse. This was a horizontal bar from which the cowboys would hang by their hands for 5-10 minutes to relieve pressure on ruptured spinal disks that came from too much bronc-fighting. Some wore eight-inch-wide heavy leather belts to keep their kidneys in place during prolonged hard rides.

When in a sense it was truly church time—when cowboys were badly injured and in need of help—they had long since learned where to go. David vividly remembers a moment in his education which was truncated when a cowboy rode up holding a bleeding hand. He had been roping a wild horse, and one of his fingers had become caught between the lariat and the saddle horn. The finger was still a part of his hand but was hanging by two tendons. His mother boiled water, sterilized a pair of surgical scissors, and scrubbed her hands and arms. With magisterial nonchalance, she "snipped the tendons, dropped the finger into the hot coals of the fire box, sewed a flap of skin over the stump, smiled sweetly, and said: 'Joe, in a month you'll never know the difference.'"

There was a pack of ferocious wolf-hounds in the country, kept by another flockmaster for the purpose of killing coyotes. The dogs seemed to relish killing rattlesnakes as well, shaking the life out of them until the festive serpents hung from the hounds' jaws like fettuccine. The ranch hand in charge of them said, "They ain't happy in the spring till they've been bit. They're used to it now, and their heads don't swell up no more." Human beings (on foot) who happened to encounter these dogs might have preferred to encounter the rattlesnakes instead. One summer afternoon, John Love was working on a woodpile when he saw two of the wolfhounds streaking down the creek in the direction of his sons, whose ages were maybe three and four. "Laddies! Run! Run to the house!" he shouted. "Here come the hounds!" The boys ran, reached the door just ahead of the dogs, and slammed it in their faces. Their mother was in the kitchen:

The hounds, not to be thwarted so easily, leaped together furiously at the kitchen windows, high above the ground. They shattered the glass of the small panes, and tried to struggle through, their front feet catching over the inside ledge of the window frame, and their heads, with slavering mouths, reaching through the broken glass. I had only time to snatch a heavy iron frying pan from the stove and face them, beating at those clutching feet and snarling heads. The terrified boys cowered behind me. The window sashes held against the onslaught of the hounds, and my blows must have daunted them. They dropped back to the ground and raced away.

In the boys' vocabulary, the word "hound" joined the word "throat" in the deep shadows, and to this day when David sees a wolfhound there is a drop in the temperature of the center of his spine.

The milieu of Love Ranch was not all wind, snow, freezing cattle, and killer dogs. There were quiet, lyrical days on end under blue, unthreatening skies. There were the redwing blackbirds on the corral fence, and the scent of moss flowers in spring. In a light breeze, the windmill turned slowly beside the wide log house, which was edged with flowers in bloom. Sometimes there were teal on the creek— and goldeneyes, pintails, mallards. When the wild hay was ready for cutting, the harvest lasted a week.

John liked to have me ride with them for the last load. Sometimes I held the reins and called "Whoa, Dan!" while the men pitched up the hay. Then while the wagon swayed slowly back over the uneven road, I lay nestled deeply beside Allan and David in the fragrant hay. The billowy white clouds moving across the wide blue sky were close, so close, it seemed there was nothing else in the universe but clouds and hay.

When the hay house was not absolutely full, the boys cleared off the dance floor of Joe Lacey's Muskrat Saloon and strapped on their roller skates. Bizarre as it may seem, there was also a Love Ranch croquet ground. And in winter the boys clamped ice skates to their shoes and flew with the wind up the creek. Alternatively, they lay down on their sleds and propelled themselves swiftly over wind-cleared, wind-polished black ice, with an anchor pin from a coyote trap in each hand. Almost every evening, with their parents, they played mah-jongg.

One fall, their mother went to Riverton, sixty-five miles away, to await the birth of Phoebe. For her sons, eleven and twelve, she left behind a carefully prepared program of study. In the weeks that followed, they were in effect enrolled in a correspondence school run by their mother. They did their French, their spelling, their arithmetic lessons, put them in envelopes, rode fifteen miles to the post office and mailed them to her. She graded the lessons and sent them back—before and after the birth of the baby.

Her hair was the color of my wedding ring. On her cheek the fingers of one hand were outspread like a small, pink starfish.

From time to time, dust would appear on the horizon, behind a figure coming toward the ranch. The boys, in their curiosity, would climb a rooftop to watch and wait as the rider covered the intervening miles. Almost everyone who went through the region stopped at Love Ranch. It had not only the sizable bunkhouse and the most capacious horse corrals in a thousand square miles but also a spring of good water. Moreover, it had Scottish hospitality—not to mention the forbidding distance to the nearest alternative cup of coffee. Soon after Mr. Love and Miss Waxham were married, Nathaniel Thomas, the Episcopal Bishop of Wyoming, came through in his Gospel Wagon, accompanied by his colleague the Reverend Theodore Sedgwick. Sedgwick later reported (in a publication called *The Spirit of Missions*):

We saw a distant building. It meant water. At this lonely ranch, in the midst of a sandy desert, we found a young woman. Her husband had gone for the day over the range. Around her neck hung a gold chain with a Phi Beta Kappa key. She was a graduate of Wellesley College, and was now a Wyoming bride. She knew her Greek and Latin, and loved her horse on the care-free prairie.

The bishop said he was searching for "heathen," and he did not linger.

Fugitive criminals stopped at the ranch fairly often. They had to—in much the way that fugitive criminals in lonely country today will sooner or later have to stop at a filling station. A lone rider arrived at the ranch one day with a big cloud of dust on the horizon behind him. The dust might as well have formed in the air the letters of the word "posse." John Love knew the rider, knew that he was wanted for murder, and knew that throughout the country the consensus was that the victim had "needed killing." The murderer asked John Love to give him five dollars, and said he would leave his pocket watch as collateral. If his offer was refused, the man said, he would find a way to take the money. The watch was as honest as the day is long. When David does his field geology, he has it in his pocket.

People like that came along with such frequency that David's mother eventually assembled a chronicle called "Murderers I Have Known." She did not publish the manuscript, or even give it much private circulation, in her regard for the sensitivities of some of the first families of Wyoming. As David would one day comment, "they were nice men, family friends, who had put away people who needed killing, and she did not wish to offend them—so many of them were such decent people."

One of these was Bill Grace. Homesteader and cowboy, he was one of the most celebrated murderers in central Wyoming, and he had served time, but people generally disagreed with the judiciary and felt that Bill, in the acts for which he was convicted, had only been "doing his civic duty." At the height of his fame, he stopped at the ranch one afternoon and stayed for dinner. Although David and Allan were young boys, they knew exactly who he was, and in his presence were struck dumb with awe. As it happened, they had come upon and dispatched a rattlesnake that day—a big one, over five feet long. Their mother decided to serve it creamed on toast for dinner. She and their father sternly instructed David and Allan not to use the word "rattlesnake" at the table. They were to refer to it as chicken, since a possibility existed that Bill Grace might not be an eater of adequate sophistication to enjoy the truth. The excitement was too much for the boys. Despite the parental injunction, gradually their conversation at the table fished its way toward the snake. Casually—while the meal was going down—the boys raised the subject of poisonous vipers, gave their estimates of the contents of local dens, told stories of snake encounters, and so forth. Finally, one of them remarked on how very good rattlers were to eat.

Bill Grace said, "By God, if anybody ever gave me rattlesnake meat I'd kill them."

The boys went into a state of catatonic paralysis. In the pure silence, their mother said, "More chicken, Bill?"

"Don't mind if I do," said Bill Grace.

Rising from the Plains

ONE YEAR, WITH DAVID LOVE, I MADE A field trip that included the Beartooth Mountains, the Yellowstone Plateau, the Hebgen earthquake zone of the Madison River, the Island Park Caldera, and parts of the Snake River Plain. Near the end of the journey, we came over Teton Pass and looked down into Jackson Hole. In a tone of sudden refreshment, he said, "Now, there is a place for a kid to cut his eye-teeth on dynamic geology."

Among others, he was referring to himself. He rode into the valley in the summer of '34. Aged twenty-one, he set up a base camp, and went off to work in the mountains. There were a number of small lakes among the Tetons at altitudes up to ten thousand five hundred feet—Cirque Lake, Mink Lake, Grizzly Bear Lake, Icefloe Lake, Snowdrift Lake, Lake Solitude—and no one knew how deep they were or how much water they might contain. The Wyoming Geological Survey wanted to know, and had offered him a summer job and a collapsible boat. He climbed the Tetons, and rowed the lakes, like Thoreau sounding depths on Walden Pond. He likes to say that the first time he was ever seasick was above timberline. If the Teton peaks were like the Alps—a transplanted segment of the Pennine Alps—there was the huge difference that just up the road from the Pennine Alps there are no geyser basins, boiling springs, bubbling muds, or lavas that froze in human time. His base camp was on Signal Mountain—by Teton standards, a hill—rising from the valley floor a thousand feet above Jackson Lake. More than fifty summers later, one day on Signal Mountain he said, "When I was a pup, I used to come up here to get away from it all."

I said, "By yourself?"

And he answered, "Oh, yes. Always. No concubines. I've always been pretty solitary. I still am."

Gouging around the mountains in his free time—and traversing the valley—he would get off his horse here and again, sit down, and think. ("You can't do geology

Hidden Lake, Grand Teton National Park, Wyoming.

in a hurry.") On horseback or on foot—from that summer forward, whenever he was there—he gathered with his eyes and his hammer details of the landscape. If he happened to come to a summit or an overlook with a wide view, he would try to spend as much of a day as possible there, gradually absorbing the country, sensing the control from its concealed and evident structure, wondering—as if it were a formal composition—how it had been done. ("It doesn't matter that I don't know what I'm looking at. Later on, it becomes clear—maybe. And maybe not. You try to put the petals back on the flower.") Some of those summits had not been visited before, but almost without exception he did not make a cairn or leave his name. ("I left my name on two peaks. When you're young and full of life, you do strange things.") Having no way to know what would or would not yield insight, he noticed almost anything. The mountain asters always faced east. Boul-

The Snake River flows below the Teton Range, Grand Teton National Park, Wyoming.

ders were far from the bedrock from which they derived. There was no quartzite in any of the surrounding mountains, but the valley was deeply filled with gold-bearing quartzite boulders. He discovered many faults in the valley floor, and failed for years to discern among them anything close to a logical sequence. There were different episodes of volcanism in two adjacent buttes. From high lookoffs he saw the barbed headwaters of streams that started flowing in one direction and then looped about and went the other way — the sort of action that might be noticed by a person carrying water on a tray. Something must have tilted this tray. From Signal Mountain he looked down at the Snake River close below, locally

sluggish and ponded, with elaborate meanders that had turned into oxbows — the classic appearance of an old river moving through low country. This was scarcely low country, and the Snake was anything but old. Several miles downstream, it took a sharp right, straightened itself out, picked up speed, and turned white. Looking down from Signal Mountain, he also noticed that moose, elk, and deer all drank from one spring just before their time of rut, crowding in, pushing and shoving to get at it ("They honk and holler and carry on"), ignoring the nearby waters of river, swamp, and lake. He named the place Aphrodisiac Spring. Over the decades, a stretch at a time, he completely circumambulated the skyline of

Teton Range from
Snake River
overlook.

Jackson Hole, camping where darkness came upon him, casting grasshoppers or Mormon crickets to catch his dinner. There were trout in the streams as big as Virginia hams. Sometimes he preferred grouse. ("I could throw a geology hammer through the air and easily knock off a blue grouse or a sage chicken. In season, of course. Hammer-throwing season. In the Absarokas, I threw at rattlesnakes, too. I don't kill rattlesnakes anymore. I've come to realize they're a part of the natural scene, and I don't want to upset it.") He carried no gun. He carries a bear bell instead. One day, when he forgot the bell, a sow grizzly stood up out of nowhere— six feet tall—and squinted at him. Suddenly, his skin felt dry and tight. ("Guess who went away.") A number of times, he

was charged by moose. He climbed a tree. On one occasion, there was no tree. He and the moose were above timberline. He happened to be on the higher ground, so he rolled boulders at the moose. One of them shattered, and sprayed the moose with shrapnel. ("The moose thought it over, and left.")
Rising from the Plains

WHERE THE STAGE ROUTE FROM CASPER to Fort Washakie had crossed a tributary of Muskrat Creek, the banks were so high and the drop to the creekbed so precipitous that the site was littered with split wagon reaches and broken wheels. Allan and David called it Jumping-Off Draw, its name on the map today. Finding numerous large bones in a meadowy bog,

Sandstone forms
on the Colorado
Plateau.

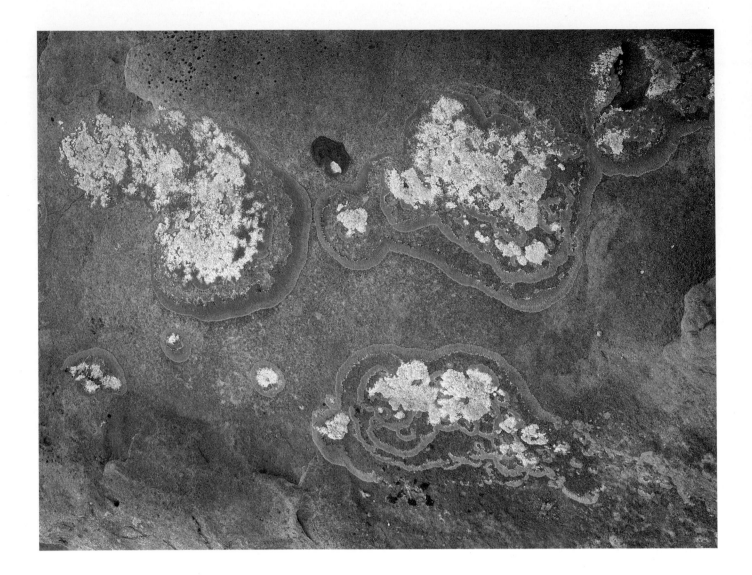

Lichen patterns near Green River, Wyoming.

they named the place Buffalo Wallows. Indians had apparently driven the bison into the swamp to kill them. One could infer that. One could also see that the swamp was there because water was bleeding from rock outcrops above the meadows. In a youth spent on horseback, there was not a lot to do but look at the landscape. The rock that was bleeding water was not just porous but permeable. It was also strong. It was the same red rock that the granary stood on, and the bunkhouse. Very evidently, it was made

of naturally cemented sand. The water could not have come from the creek. The Buffalo Wallows were sixty feet higher than the creek. The sandstone layers tilted north. They therefore reached out to the east and west. There was high ground to the east. The water must be coming down from there. One did not need a Ph.D. from Yale to figure that out—especially if one was growing up in a place where so much rock was exposed. Pending further study, his interpretation of the Buffalo Wallows was just a horseback guess. All through his life, when he would make a shrewd surmise he would call it a horse-back guess.

The water in the sandstone produced not only the bogs but the adjacent meadows as well—in this otherwise desiccated terrain. From the meadows came hay. There was an obvious and close relationship between bedrock geology and ranching. David would not have articulated that in just those words, of course, but he thought about the subject much of the time, and he was drawn to be a geologist in much the way that someone growing up in Gloucester, Massachusetts, would be drawn to be a fisherman. "It was something to think about on long rides day after day when everything was so monotonous," he remarked not long ago. "Monotony was what we fought out there. Day after day, you had nothing but the terrain around you—you had nothing to think about but why the shale had stripes on it, why the boggy places were boggy, why the vegetation grew where it did, why trees grew only on certain types of rock, why water was good in some places and bad in others, why the meadows were where they were, why some creek crossings were so sandy they were all but impassable. These things were very real, very practical. If you're in bedrock, caliche, or gumbo, the going is hard. Caliche is lime precipitate at the water table—you learn some geology the hard way. There was nothing else to be interested in. Everything depended on geology. Any damn fool could see that the vegetation was directly responsive to the bedrock. Hence birds and wildlife were responsive to it. We were responsive to it. In winter, our life was governed by where the wind blew, where snow accumulated. We could see that these natural phenomena were not random—that

they were controlled, that there was a system. The processes of erosion and deposition were things we grew up with. An insulated society does not see how important terrain is to someone who has to understand it in order to live with it. Much of it meant life or death for the animals, and therefore survival for us. If there was one thing we learned, it was that you don't fight nature. You live with it. And you make the accommodations—because nature does not accommodate."
Basin and Range

Eroded lands near South Pass, Wyoming.

IN AGE, THE BLUE STONE APPROACHED five hundred million years. Captain Howard Stansbury, USA, whose name would rest upon the mountains of which the rock was a component, was approaching fifty when he came into the Great Basin in 1849. He had been making lighthouses in Florida. The government preferred that he survey the salt lake. With sixteen mules, a water keg, and some India-rubber bags, he went around the lake, and then some. People told him not to try it. He ran out of water but not of luck. And he came back with the story of having seen—far out on the westward flats—scattered books, clothing, trunks, tools, chains, yokes, dead oxen, and abandoned wagons. The Donner Party went around the nose of the Stansburys in late August, 1846, rock on their left, lake marshes on their right. This huge blue roadcut, in its supranatural way, would have frightened them to death. They must have filed along just about where Deffeyes had parked the pickup, on the outside shoulder of the interstate. Deffeyes and I went back across the road, waiting first for a three-unit seven-axle tractor-trailer to pass. Deffeyes described it as "a freaking train."

Stansbury Mountains, Skull Valley . . . The Donner Party found good grass in Skull Valley, and good water, and a note by a post at a spring. It had been torn to shreds by birds. The emigrants pieced it together. "Two days—two nights—hard driving—cross desert—reach water." They went out of Skull Valley over the Cedar Mountains into Ripple Valley and over Grayback Mountain to the Great Salt Lake Desert. Grayback Mountain was basalt, like the Watchungs of New Jersey. The New Jersey basalt flowed about two hundred million years ago. The Grayback Mountain basalt flowed thirty-eight mil-

lion years ago. Well into this century, it was possible to find among the dark-gray outcrops of Grayback Mountain pieces of wagons and of oxhorn, discarded earthenware jugs. The snow suddenly gone now, and in cold sunshine, Deffeyes and I passed Grayback Mountain and then had the Great Salt Lake Desert before us—the dry bed of Bonneville—broader than the periphery of vision. The interstate runs close to but not parallel to the wagon trail, which trends a little more northwesterly. The wagon trail aims directly at Pilot Peak of the Pilot Range, which we could see clearly, upward of fifty miles away—a pyramidal summit with cloud coming off it in the wind like a banner unfurling. Across the dry lakebed, the emigrants homed on Pilot Peak, standing in what is now Nevada, above ten thousand feet. Along the fault scarp, at the base of Pilot Peak, are cold springs. When the emigrants arrived at the springs, their tongues were bloody and black.

"Imagine those poor sons of bitches out here with their animals, getting thirsty," Deffeyes said. "It's a wonder they didn't string the guy that invented this route up by his thumbs."

The flats for the most part were alkaline, a leather-colored mud superficially dry. Dig down two inches and it was damp and greasy. Come a little rain and an ox could go in to its knees. The emigrants made no intended stops on the Great Salt Lake Desert. They drove day and night for the Pilot Range. In the day, they saw mirages—towers and towns and

Pioneer graves, Skull Valley, Utah.

shimmering lakes. Sometimes the lakes were real—playa lakes, temporary waters after a storm. Under a wind, playa lakes move like puddles of mercury in motion on a floor—two or three hundred square miles of water on the move, here today, there tomorrow, gone before long like a mirage, leaving wagons mired in unimagined mud. Very few emigrants chose to cross the Bonneville flats, although the route was promoted as a shortcut—"a nigher route"—rejoining the main migration four basins into Nevada. It was the invention of Lansford Hastings and was known as the Hastings Cutoff. Hastings wrote the helpful note in Skull Valley. His route was geologically unfavorable, but this escaped his knowledge and notice. His preoccupations were with politics. He wished to become President of California. He saw California—for the moment undefendably Mexican—as a new nation, under God, conceived at liberty and dedicated to the proposition that anything can be accomplished through promotion: President Lansford Hastings, in residence in a Western White House. His strategy for achieving high office was to create a new shortcut on the way west, to promote both the route and the destination through recruiting and pamphleteering, to attract emigrants by the thousands year after year, and as their counsellor and deliverer to use them as constituent soldiers in the promised heaven. He camped beside the trail farther east. He attracted the Donners. He attracted Reeds, Kesebergs, Murphys, McCutchens, drew them southward away from the main trek and into the detentive

scrub oak made fertile by the limestones of the Wasatch. The Donners were straight off the craton—solid and trusting, from Springfield, Illinois. Weeks were used hacking a path through the scrub oaks, which were living barbed wire. Equipment was abandoned on the Bonneville flats to lighten up loads in the race against thirst. Even in miles, the nigher route proved longer than the one it was shortcutting, on the way to a sierra that was named for snow.

Basin and Range

Bizarre salt patterns, Bonneville Salt Flats, Utah.

3.   THE RIVER

ANYONE INTERESTED, FOR whatever reason, in the study of water in the West will in the end concentrate on the Colorado, wildest of rivers, foaming, raging, rushing southward—erratic, head-long, incongruous in the desert. The Snake, the Salmon, the upper Hudson—all the other celebrated white torrents—are not in the conversation if the topic is the Colorado. This is still true, although recently (recently in the long span of things, actually within the past forty years) the Colorado has in places been subdued. The country around it is so dry that Dominy's county in Wyoming is a rain forest by comparison. The states of the basin need water, and the Colorado is where the water is. The familiar story of contention for water rights in the Old West—Alan Ladd shooting it out with Jack Palance over some rivulet God knows where—has its mother narrative in the old and continuing story of rights to the waters of the Colorado. The central document is something called the Colorado River Compact, in which the basin is divided in two, at a point close to the Utah-Arizona line. The states of the Upper Basin are allowed to take so much per year. The Lower Basin gets approxi-

mately an equal share. And something gratuitous is passed on to Mexico. The Colorado lights and slakes Los Angeles. It irrigates Arizona. The odd thing about it is that all its writhings and foamings and spectacular rapids lead to nothing. The river rises in the Rockies, thunders through the canyons, and is so used by mankind that when it reaches the Gulf of California, fourteen hundred miles from its source, it literally trickles into the sea. *Encounters with the Archdruid*

IN THE VIEW OF CONSERVATIONISTS, there is something special about dams, something—as conservation problems go—that is disproportionately and metaphysically sinister. The outermost circle of the Devil's world seems to be a moat filled mainly with DDT. Next to it is a moat of burning gasoline. Within that is a ring of pinheads each covered with a million people—and so on past phalanxed bulldozers and bicuspid chain saws into the absolute epicenter of Hell on earth, where stands a dam. The implications of the dam exceed its true level in the scale of environmental catastrophes. Conserva-tionists who can hold themselves in reasonable check before new oil spills and

Colorado River in winter.

fresh megalopolises mysteriously go insane at even the thought of a dam. The conservation movement is a mystical and religious force, and possibly the reaction to dams is so violent because rivers are the ultimate metaphors of existence, and dams destroy rivers. Humiliating nature, a dam is evil—placed and solid.

"I hate all dams, large and small," David Brower informs an audience.

A voice from the back of the room asks, "Why are you conservationists always against things?"

"If you are against something, you are for something," Brower answers. "If you are against a dam, you are for a river."

Encounters with the Archdruid

WHAT SEEMED UNIMAGINABLE BESIDE the river in the canyon was that all that wild water had been processed, like pork slurry in a hot-dog plant, upstream in the lightless penstocks of a big dam. Perspective is where you find it, though, and with this in mind Dominy had taken Brower and me, some days earlier, down into the interior of his indisputable masterpiece, the ten-million-ton plug in Glen Canyon. We had seen it first from the air and then from the rim of Glen Canyon, and the dam had appeared from on high to be frail and surprisingly small, a gracefully curving wafer wedged flippantly into the river gorge, with a boulevard of blue water on one side of it and a trail of green river on the other. No national frontier that I can think of separates two worlds more dissimilar than the reservoir and the river. This frontier has a kind of *douane* as well, administered by men who work in a perfectly circular room deep inside the dam. They wear slim ties and white short-sleeved shirts. They make notes on clipboards. They sit at desks, and all around them, emplaced in the walls of the room,

are gauges and dials, and more gauges and dials. To get to this control room, we rode about five hundred feet down into the dam in an elevator, and as we descended Dominy said, "People talk about environment. We're doing something about it." His eyes gleamed with humor. He led us down a long passageway and through a steel door. The men inside stood up. From the devotional look in their eyes, one might have thought that Marc Mitscher had just walked into the engine room of the carrier *Lexington* on the night after the Battle of the Philippine Sea. This was, after all, the man they called the Kmish. Throughout Reclamation, Dominy was known as the Kmish. Standing there, he introduced each man by name. He asked the elevation of Lake Powell.

"Three thousand five hundred and seventy-seven point two zero feet, sir."

Dominy nodded. He was pleased. When the level of the surface is lowered, a distinct band, known to conservationists as "the bathtub ring," appears along the cliff faces that hold the reservoir. Three thousand five hundred and seventy-seven point two zero would eliminate that, and a good thing, too, for on this day—one hundred years to the sunrise since the day Major Powell reached Glen Canyon on his first expedition—Lake Powell was to be dedicated.

"What are we releasing?" Dominy asked.

"Four thousand three hundred and fifty-six point zero cubic feet per second, sir."

"That's about normal," Dominy said. "Just a little low."

At their consoles, turning knobs, flicking switches, the men in the control room continually create the river below the dam. At that moment, they were releasing something like fourteen hundred

tons of water every ten seconds—or, in their terminology, one acre-foot.

"We have eight generating units," Dominy went on. "When we want to make peaking power, we turn them up full and send a wall of water downstream. The rubber rafts operate with licenses, and the guides know the schedule of releases."

Dominy then took us all the way down—down in another elevator, down concrete and spiral stairways, along ever-deeper passageways and down more

stairways—until we were under the original bed of the Colorado and at the absolute bottom of the dam, seven hundred and ten feet below the crest. "I don't want Dave Brower to be able to say he didn't see everything," Dominy said—and I could not help admiring him for it, because the milieu he had taken us into could easily be misunderstood. Water was everywhere. Water poured down the spiral staircases. It streamed through the passageways. It fell from the ceilings. It ran

Gunsight Butte and Navajo Mountain in Lake Powell, Glen Canyon National Recreation Area, Utah.

from the walls. In some places, sheets of polyethylene had been taped to the concrete. At the bottom, Glen Canyon Dam is three hundred feet thick, but nearly two hundred miles of reservoir was pressing against it, and it had cracked. The Colorado was pouring through. "We may have to get some Dutch boys in here with their thumbs," Dominy said. "The dam is still curing. It hasn't matured yet. So we aren't doing much of anything about this now. We will soon. We have a reinjectionable grouting system; it's an idea I picked up in Switzerland. The crack water is declining anyway. The crack may be sealing itself. It's not serious. You just cannot completely stop the Colorado River."

Encounters with the Archdruid

SEVEN YEARS EARLIER, WE COULD HAVE flown north through Glen Canyon at an altitude of four hundred feet over the riverbed, and that, in a way, is what we did now. We got into a nineteen-foot gray boat—its hull molded for speed, a Buick V-6 engine packed away somewhere, a two-way radio, and the black-lettered words *United States Government* across the stern—and up the lake we went at twenty knots, for three days spraying arches of clear water toward red-and-black-streaked tapestry walls, pinnacle spires, and monument buttes. The Utah canyonland had been severed halfway up by a blue geometric plane, creating a waterscape of interrupted shapes, spectacularly unnatural, spectacularly beautiful. If we stopped for lunch, nudging up to a cool shadowing wall, we were in fact four hundred feet up the sheer side of what had been an immense cliff above the river, and was still an immense cliff—Wingate, Kayenta, Navajo Sandstones—above the

lake. The boat sped on among hemispherical islands that had once been mountainous domes. It wheeled into Caprian bays. Arched overhangs formed grottoes in what had once been the lofty ceilings of natural amphitheatres.

Above the sound of the engine, Dominy shouted, "Who but Dominy would build a lake in the desert? Look at the country around here! No vegetation. No precipitation. It's just not the setting for a lake under any natural circumstances. Yet it is the most beautiful lake in the world."

"A thousand people a year times ten thousand years times ten thousand years will never see what was there," Brower said. He pointed straight down into the water. Then he opened a can of beer. The beer was in a big container full of ice. The ice had been made from water of the reservoir—reclaimed pellets of the Colorado. The container held dozens of cans of beer and soft drinks, enough for ten men anywhere else, but even on the lake the air was as dry as paper and the sun was a desert sun, and we held those cans in the air like plasma, one after another, all day long. Brower, the aesthetician, likes beer cans. Not for him are the simple biases of his throng. He really appreciates the cans themselves—their cylindrical simplicity, their beautifully crafted lithography. Brower's love of beauty is so powerful it leaps. It sometimes lands in unexpected places. Looking out over the lake at canyon walls flashing in reflected light, he slowly turned his Budweiser in his hand, sipped a little, and then said, "Lake Powell does not exist. I have never seen anything like it before. It's an incredibly beautiful reservoir. It must be the most beautiful reservoir in the world. I just wish you could hold the water level where it is now, Floyd."

Dominy smiled. The lake would become more and more beautiful as it continued to fill, he said. It would go up another hundred and twenty feet, revising vistas as it rose, and the last thirty-five feet would be the most dramatic, because the water at that elevation would reach far into the canyonland.

"You can't duplicate this experience— this lake—anywhere else," Brower said. "But neither can you enjoy the original experience. That's the trouble. I camped under here once. It was a beautiful camp-site. The river was one unending camp-site. The ibis, the egrets, the wild blue herons are gone. Their habitat is gone— the mudbanks along the river."

"We've covered up a lot of nice stuff, there's no question about that, but you've got to admit that as far as views are concerned we've opened up a lot. Look. You can see mountains."

"The Henry Mountains," Brower said. "They were the last mountain range discovered in the lower forty-eight."

For my part, I kept waiting to see the

Henry Mountains, Utah.

lake. "Lake," as I sensed the word, called to mind a fairly compact water-filled depression in high terrain, with bends and bays perhaps obscuring some parts from others, but with a discernible center, a middle, a place that was farther from shore than any other, and from which a sweeping view of shoreline could be had in all directions. This was a provincialism, based on a Saranac, a Sunapee, a Mooselookmeguntic, and it had left me unprepared for Lake Powell, a map of which looks like a diagram of the human nervous system. The deep spinal channel of Glen Canyon, which was once the path of the Colorado, is now the least interesting part of Lake Powell. The long, narrow bays that reach far into hundreds of tributary canyons are the absorbing places to enter—the boat rounding bends between ever-narrowing walls among reflections of extraordinary beauty on wind-slickened rock. These were the places—these unimaginably deep clefts in the sandstone—that most stirred and most saddened Brower, who remembered wading through clear pools under cottonwood trees four hundred feet below the arbitrary level on which we floated.

In Face Canyon, the boat idled slowly and moved almost silently through still water along bending corridors of rock. "There used to be pools and trees in this little canyon," Brower said. "Cottonwoods, willows."

"Poison ivy, jimsonweed," Dominy said.

"Little parks with grasses. Water always running," Brower went on.

The rock, dark with the oxidation known as desert varnish, appeared to be a rich blue. Desert varnish somehow picks up color from the sky. The notes of a canyon wren descended the pentatonic scale. "That's the music here—the best there is," Brower said. "There used to be paper shells of surface mud on the floor of this canyon, cracking, peeling. Damn it, that was handsome."

"On balance, I can't lament what's been covered up," Dominy said.

In Cascade Canyon, on a ledge that had once been hundreds of feet high, grew a colony of mosses and ferns. "Now there's a hanging garden that's going to get water beyond its wildest dreams," Dominy said. "But unfortunately, like welfare, the water is going to drown it."

In Brower's memory, the most beautiful place in all the region of Glen Canyon was a cavernous space, under vaulting rock walls, that had been named the Cathedral in the Desert. The great walls arched toward one another, forming high and almost symmetrical overlapping parabolas. They enclosed about an acre of ground, in which had grown willows, grasses, columbine, and maidenhair fern. The center of this scene was a slim waterfall, no more than a foot in diameter, that fell sixty feet into a deep and foaming pool. From it a clear stream had flowed through the nave and out to the Colorado. The government boat now entered the Cathedral. Dominy switched off the engine. Water was halfway to the ceiling, and the waterfall was about ten feet high. It was cool in there, and truly beautiful—the vaulted ceiling, the sound of the falling water, the dancing and prismatic reflections, the echo of whispers. It had been beautiful in there before the reservoir came, and it would continue to be so, in successive stages, until water closed the room altogether.

Encounters with the Archdruid

Varnished wall in Horseshoe Canyon, Canyonlands National Park, Utah.

**Rainbow Bridge,
Rainbow Bridge
National
Monument,
Utah.**

reach into the deep groove of the creek bed below Rainbow Bridge and will fill it to a level twenty-two feet below the base of the span. To Brower, this is simple sacrilege. To Dominy, it is a curious and agreeable coincidence that the water will stop just there.
Encounters with the Archdruid

THE RAFT CONSISTS OF, AMONG OTHER things, two neoprene bananas ten yards long. These pontoons, lashed to a central rubber barge, give the over-all rig both lateral and longitudinal flexibility. The river sometimes leaps straight up through the raft, but that is a mark of stability rather than imminent disaster. The raft is informal and extremely plastic. Its lack of rigidity makes it safe.

This is isolation wilderness: two or three trails in two hundred miles, otherwise no way out but down the river with the raft. Having seen the canyon from this perspective, I would not much want to experience it another way. Once in a rare while, we glimpse the rims. They are a mile above us and, in places, twelve miles apart. All the flat shelves of color beneath them return the eye by steps to the earliest beginnings of the world—from the high white limestones and maroon Hermit Shales of Permian time to the red sandstones that formed when the first reptiles lived and the vermilion cliffs that stood contemporary with the earliest trees. This Redwall Limestone, five hundred feet thick, is so vulnerable to the infiltrations of groundwater that it has been shaped, in the seas of air between the canyon rims, into red towers and red buttes, pillars, caverns, arches, and caves. The groundwater runs for hundreds of miles between the layers of that apparently bone-dry desert rock and bursts out into the canyon in stepped cascades or ribbon falls. We are

RAINBOW BRIDGE WAS FORMED—IN AN era when the land was uplifting—by waters that raced off Navajo Mountain and punched through a sandstone wall. Pushing gravel and boulders through the opening, cutting down and cutting wide, the creek, in centuries, made the gigantic stone span that crosses it now. Thick and red, immense against the sky, it would fit over the national Capitol dome. It is the largest known natural bridge on earth. When Lake Powell is full, still water will

looking at such a waterfall right now, veiling away from the Redwall, high above us. There is green limestone behind the waterfall, and pink limestone that was pressed into being by the crushing weight of the ocean at the exact time the ocean itself was first giving up life — amphibious life — to dry land. Beneath the pink and green limestones are green-gray shales and dark-brown sandstones — Bright Angel Shale, Tapeats Sandstone — that formed under the fathoms that held the first general abundance of marine life. Tapeats Sea was the sea that compressed the rock that was cut by the river to create the canyon. The Tapeats Sandstone is the earliest rock from the Paleozoic Era, and beneath it the mind is drawn back to the center of things, the center of the canyon, the cutting plane, the Colorado. Flanked by its Bass Limestones, its Hotauta Conglomerates, its Vishnu Schists and Zoroaster Granites, it races in white water through a Pre-cambrian here and now.

The Maze, Canyonlands National Park, Utah.

Entrada standstone
of the Fiery
Furnace, Arches
National Park,
Utah.

The river has worked its way down into the stillness of original time.

Brower braces his legs and grips one of the safety ropes that run along the pontoons. He says, "How good it is to hear a living river! You can almost hear it cutting."

Dominy pulls his Lake Powell hat down firmly around his ears. He has heard this sort of thing before. Brower is sug-

gesting that the Colorado is even now making an ever deeper and grander Grand Canyon, and what sacrilege it would be to dam the river and stop that hallowed process. Dominy says, "I think most people agree, Dave, that it wasn't a river of this magnitude that cut the Grand Canyon."

Brower is too interested in the coming rapid to respond. In this corridor of calm, we can hear the rapid ahead. Rapids and waterfalls ordinarily take shape when rivers cut against resistant rock and then come to a kind of rock that gives way more easily. This is not the case in the Grand Canyon, where rapids occur beside the mouths of tributary creeks. Although these little streams may be dry much of the year, they are so steep that when they run they are able to fling considerable debris into the Colorado—sand, gravel, stones, rocks, boulders. The debris forms dams, and water rises upstream. The river is unusually quiet there—a lakelike quiet—and then it flows over the debris, falling suddenly, pounding and crashing

The Colorado River flows below the rim at Dead Horse Point State Park, Utah.

Badger Rapid of the Colorado River, Grand Canyon National Park.

through the boulders. These are the rapids of the Grand Canyon, and there are a hundred and sixty-one of them. Some have appeared quite suddenly. In 1966, an extraordinarily heavy rain fell in a small area of the north rim, and a flash flood went down Crystal Creek, dumping hundreds of tons of rock into the river at Mile 99. This instantly created the Crystal Rapids, one of the major drops in the Colorado. In rare instances—such as the rapid we are now approaching—the river has exposed resistant Pre-cambrian rock

Colorado River in eerie storm light, Grand Canyon National Park, Arizona.

that contributes something to the precipitousness of the flow of white water. The roar is quite close now. The standing waves look like blocks of cement. Dominy emits a cowboy's yell. My notes go into a rubber bag that is tied with a string. This is the Bedrock Rapid.
Encounters with the Archdruid

THERE IS SOMETHING QUITE DECEPTIVE in the sense of acceleration that comes just before a rapid. The word "rapid" itself is, in a way, a misnomer. It refers only to the speed of the white water relative to the speed of the smooth water that leads into and away from the rapid. The white water is faster, but it is hardly "rapid." The Colorado, smooth, flows about seven

miles per hour, and, white, it goes perhaps fifteen or, at its whitest and wildest, twenty miles per hour—not very rapid by the standards of the twentieth century. Force of suggestion creates a false expectation. The mere appearance of the river going over those boulders—the smoky spray, the scissoring waves—is enough to imply a rush to fatality, and this endorses the word used to describe it. You feel as if you are about to be sucked into some sort of invisible pneumatic tube and shot like a bullet into the dim beyond. But the white water, though faster than the rest of the river, is categorically slow. Running the rapids in the Colorado is a series of brief experiences, because the rapids themselves are short. In them, with the raft folding and bending—sudden hills of water filling the immediate skyline—things happen in slow motion. The projector of your own existence slows way down, and you dive as in a dream, and gradually rise, and fall again. The raft shudders across the ridgelines of water cordilleras to crash softly into the valleys beyond. Space and time in there are something other than they are out here. Tents of water form overhead, to break apart in rags. Elapsed stopwatch time has no meaning at all.

Encounters with the Archdruid

BROWER AND I WENT INTO THE STREAM and into the cliff. The current was not powerful, coming through the rock, and the water was only four feet deep. I swam, by choice—the water felt so good. It felt

Soap Creek Rapid of the Colorado River.

cool, but it must have been about seventy-five degrees. It was cooler than the air. Within the cliff was deep twilight, and the echoing sound of the moving water. A bend to the right, a bend to the left, right, left—this stone labyrinth with a crystal stream in it was moment enough, no matter where it ended, but there lay beyond it a world that humbled the mind's eye. The walls widened first into a cascaded gorge and then flared out to become the ovate sides of a deep valley, into which the stream rose in tiers of pools and waterfalls. Some of the falls were only two feet high, others four feet, six feet. There were hundreds of them. The pools were as much as fifteen feet deep, and the water in them was white where it plunged and foamed, then blue in a wide circle around the plunge point, and pale green in the outer peripheries. This was Havasu Canyon, the immemorial home of the Havasupai, whose tribal name means "the people of the blue-green waters." We climbed from one pool to another, and swam across the pools, and let the waterfalls beat down around our shoulders. Mile after mile, the pools and waterfalls continued. The high walls of the valley were bright red. Nothing grew on these dry and flaky slopes from the mesa rim down about two-thirds of the way; then life began to show in isolated barrel cactus and prickly pear. The cacti thickened farther down, and below them was riverine vegetation—green groves of oak and cottonwood, willows and tamarisk, stands of cattail, tall grasses, moss, watercress, and maidenhair fern. The Havasupai have lived in this place for hundreds, possibly thousands, of years, and their population has remained stable. There are something like two hundred of them. They gather nuts on the canyon rim in winter and grow vegetables in the canyon in summer.

Limestone walls above Havasu Creek, Grand Canyon National Park.

They live about twelve miles up Havasu Creek from the Colorado. Moss covered the rocks around the blue-and-green pools. The moss on dry rock was soft and dense, and felt like broadloom underfoot. Moss also grew below the water's surface, where it was coated with travertine, and resembled coral. The stream was loaded with calcium, and this was the physical explanation of the great beauty of Havasu Canyon, for it was the travertine — crystalline calcium carbonate — that had both fashioned and secured the all but unending stairway of falls and pools. At the downstream lip of each plunge pool, calcium deposits had built up into natural dams, and these travertine dams were what kept Havasu Creek from running freely downhill. The dams were whitish tan, and so smooth and symmetrical that they might have been finished by a mason. They were two or three feet high. They sloped. Their crests were flat and smooth and with astonishing uniformity were about four inches thick from bank to bank. Brower looked up at the red canyon walls. He was sitting on the travertine, with one foot in a waterfall, and I was treading the green water below him. He said, "If Hualapai Dam had been built, or were ever built, this place where you are swimming would be at the bottom of a hundred feet of water." It was time to go back to the Colorado. I swam to the travertine dam at the foot of the pool, climbed up on it and dived into the pool below it, and swam across and dived again, and swam and dived—and so on for nearly two miles. Dominy was waiting below. "It's fabulous," he said. "I know every river canyon in the country, and this is the prettiest in the West."

Encounters with the Archdruid

MILE 177, 9:45 A.M. THE WATER IS QUITE deep and serene here, backed up from the rapid. Lava Falls is two miles downstream, but we have long since entered its chamber of quiet.

"The calm before the storm," Brower says.

The walls of the canyon are black with lava—flows, cascades of lava. Lava once poured into the canyon in this segment of the river. The river was here, much in its present form. It had long since excavated the canyon, for the volcanism occurred in relatively recent time. Lava came up through the riverbed, out from the canyon walls, and even down over the rims. It sent the Colorado up in clouds. It hardened, and it formed a dam and backed water two hundred miles.

"If a lava flow were to occur in the Grand Canyon today, Brower and the nature lovers would shout to high heaven that a great thing had happened," Dominy said, addressing everyone in the raft. "But if a man builds a dam to bring water and power to other men, it is called desecration. Am I right or wrong, Dave? Be honest."

"The lava dam of Quaternary time was eventually broken down by the river. This is what the Colorado will do to the Dominy dams that are in it now or are ever built. It will wipe them out, recover its grade, and go on about its business. But by then our civilization and several others will be long gone."

We drift past an enormous black megalith standing in the river. For eighty years, it was called the Niggerhead. It is the neck of a volcano, and it is now called Vulcan's Forge. We have a mile to go. Brower talks about the amazing size of the crystals on the canyon walls, the morning light in the canyon, the high palisades of columnar basalt. No one else says much of anything. All jokes have been cracked twice. We are just waiting, and the first thing we hear is the sound. It is a big, tympanic sound that increasingly fills the canyon. The water around us is dark-green glass. Five hundred yards. There it is. Lava Falls. It is, of course, a rapid, not a waterfall. There is no smooth lip. What we now see ahead of us at this distance appears to be a low whitewashed wall.

The raft touches the riverbank. Sanderson gets out to inspect the rapid, and we go, too. We stand on a black ledge, in the roar of the torrent, and look at the

water. It goes everywhere. From bank to bank, the river is filled with boulders, and the water smashes into them, sends up auroras of spray, curls thickly, and pounds straight down into bomb-crater holes. It eddies into pockets of lethal calm and it doubles back to hit itself. The drop is prodigious—twenty-six feet in a hundred yards—but that is only half the story. Prospect Creek, rising black-walled like a coal chute across the river, has shoved enough rock in here to stop six rivers, and this has produced the preëminent rapid of the Colorado.

When Dominy stepped up on the ledge and into the immediacy of Lava Falls, he shouted above the thunder, "Boy, that's a son of a bitch! Look at those *rocks!* See that hole over there? Jesus! Look at that one!"

Brower said, "Look at the way the water swirls. It's alive!"

The phys.-cd. teacher said, "Boy, that could tear the hell out of your bod."

Brower said, "Few come, but thousands drown."

Dominy said, "If I were Jerry, I'd go to the left and then try to move to the right."

Lava protruded from the banks in jagged masses, particularly on the right, and there was a boulder there that looked like an axe blade. Brower said, "I'd go in on the right and out on the left."

My own view was that the river would make all the decisions. I asked Sanderson how he planned to approach what we saw there.

"There's only one way to do it," he said. "We go to the right."

Beach reflections on the Colorado River.

The raft moved into the river slowly, and turned, and moved toward the low white wall. A hundred yards. Seventy-five yards. Fifty yards. It seems odd, but I did not notice until just then that Brower was on the raft. He was, in fact, beside me. His legs were braced, his hands were tight on a safety rope, and his Sierra Club cup was hooked in his belt. The tendons in his neck were taut. His chin was up. His eyes looked straight down the river. From a shirt pocket Dominy withdrew a cigar. He lighted it and took a voluminous drag. We had remaining about fifteen seconds of calm water. He said, "I might bite an inch off the end, but I doubt it." Then we went into Lava Falls.

Water welled up like a cushion against the big boulder on the right, and the raft went straight into it, but the pillow of crashing water was so thick that it acted on the raft like a great rubber fender between a wharf and a ship. We slid off the rock and to the left—into the crater-scape. The raft bent like a V, flipped open, and shuddered forward. The little outboard—it represented all the choice we had—cavitated, and screamed in the air. Water rose up in tons through the bottom of the raft. It came in from the left, the right, and above. It felt great. It covered us, pounded us, lifted us, and heaved us scudding to the base of the rapid.

For a moment, we sat quietly in the calm, looking back. Then Brower said, "The foot of Lava Falls would be two hundred and twenty-five feet beneath the surface of Lake Dominy."

Dominy said nothing. He just sat there, drawing on a wet, dead cigar. Ten minutes later, however, in the dry and baking Arizona air, he struck a match and lighted the cigar again.

Encounters with the Archdruid

4. MOUNTAINS, LAKES, WINDS

THE GREAT SALT LAKE REACHED out to our right and disappeared in snow. In a sense, there was no beach. The basin flatness just ran to the lake and kept on going, wet. The angle formed at the shoreline appeared to be about 179.9 degrees. There were dark shapes of islands, firmaments in the swirling snow—elongate, north-south-trending islands, the engulfed summits of buried ranges. "Chemically, this is one of the toughest environments in the world," Deffeyes said. "You swing from the saltiest to the most dilute waters on the planet in a matter of hours. Some of the most primitive things living are all that can take that. The brine is nearly saturated with sodium chloride. For a short period each year, so much water comes down out of the Wasatch that large parts of the lake surface are relatively fresh. Any creature living there gets an osmotic shock that amounts to hundreds of pounds per square inch. No higher plants can take that, no higher animals—no multicelled organisms. Few bacteria. Few algae. Brine shrimp, which do live there, die by the millions from the shock."

I have seen the salt lake incredibly beautiful in winter dusk under snow-streamer curtains of cloud moving fast through the sky, with the wall of the Wasatch a deep rose and the lake islands rising from what seemed to be rippled slate. All of that was now implied by the mysterious shapes in the foreshortening snow. I didn't mind the snow. One June day, moreover, with Karen Kleinspehn—on her way west for summer field work—I stopped in the Wasatch for a picnic of fruit and cheese beside a clear Pyrenean stream rushing white over cobbles of quartzite and sandstone through a green upland meadow—cattle in the meadow, cottonwoods along the banks of this clear, fresh, suggestively confident, vitally ignorant river, talking so profusely on its way to its fate, which was to move among paradisal mountain landscapes until, through a terminal canyon, the Great Basin drew it in. No outlet. Three such rivers feed the Great Salt Lake. It does indeed consume them. Descending, we ourselves went through a canyon so narrow that the Union Pacific Railroad was in the median of the interstate and on into an even steeper canyon laid out as if for skiing in a hypnotizing rhythm of christiania turns under high walls of rose-brick

Storm waves on Great Salt Lake.

Great Salt Lake
seen from the
Wasatch Mountains,
Utah.

Nugget sandstone and brittle shattered marine limestone covered with scrub oaks. "Good God, we are dropping out of the sky," said Kleinspehn, hands on the wheel, plunging through the big sheer roadcuts, one of which suddenly opened to distance, presented the Basin and Range.

" 'This is the place.' "

"You can imagine how he felt."

In the foreground was the alabaster city, with its expensive neighborhoods strung out along the Wasatch Fault, getting ready to jump fifteen feet. In the distance were the Oquirrhs, the Stansburys, the lake. Sunday afternoon and the Mormons were out on the flats by the water in folding chairs at collapsible tables, end to end like refectory tables, twenty people down to dinner, with acres of beach-flat all to themselves and seagulls around them like sacred cows. To go swimming, we had to walk first—several hundred yards straight out, until the water was ankle-deep. Then we lay down on our backs and floated. I have never been able to float. When I took the Red Cross tests, age nine to fifteen, my feet went down and I hung in the water with my chin wrenched up like something off Owl

Big Cottonwood Canyon of the Wasatch Mountains, Utah.

Creek Bridge. I kicked, slyly kicked to push my mouth above the surface and breathe. I could not truly float. Now I tried a backstroke and, like some sort of hydrofoil, went a couple of thousand feet on out over the lake. Only my heels, rump, and shoulder blades seemed to be wet. I rolled over and crawled. I could all but crawl on my hands and knees. And this was June, at the south end — the least salty season, the least salty place in the whole of the Great Salt Lake.

Basin and Range

A CLIMBER'S GUIDE TO THE HIGH SIERRA (Sierra Club, 1954) lists thirty-three peaks in the Sierra Nevada that were first ascended by David Brower. "*Arrowhead.* First ascent September 5, 1937, by David R. Brower and Richard M. Leonard. . . . *Glacier Point.* First ascent May 28, 1939, by Raffi Bedayan, David R. Brower, and Richard M. Leonard. . . . *Lost Brother.* First ascent July 27, 1941, by David R. Brower. . . ." Brower has climbed all the Sierra peaks that are higher than fourteen thousand feet. He once started out at midnight, scaled the summit of Mount Tyndall (14,025) by 3 A.M., reached the summit of Mount Williamson (14,384) by 7 A.M., and was on top of Mount Barnard (14,003) at noon. He ate his lunch — nuts, raisins, dried apricots — and he went to sleep. He often went to sleep on the high peaks. Or he hunted around for ice, removing it in wedges from cracks in the granite, sucking it to slake his thirst. If it was a nice day, he would stay put for as

Granite slopes
leading to Half
Dome in Yosemite
National Park,
Sierra Nevada,
California.

much as an hour and a half. "The summit is the anticlimax," he says. "The way up is the thing. There is a moment when you know you have the mountain by the tail. You figure out how the various elements go together. You thread the route in your mind's eye, after hunting and selecting, and hitting dead ends. Finally, God is good enough. He built the mountain right, after all. A pleasant surprise. If you don't make it and have to go back, you play it over and over again in your mind.

Maybe this would work, or that. Several months, a year, or two years later, you do it again." When Brower first tried to climb the Vazquez Monolith, in Pinnacles National Monument, he was stopped cold, as had been every other climber ever, for the face of the monolith was so smooth that Brower couldn't even get off the ground. Eventually, someone else figured out how to do that, but, as it happened, was stopped far shy of the summit. When Brower heard about this, he went to his typewriter, wrote a note identifying himself as the first man to ascend Vazquez Monolith, and slipped the note into a small brass tube. In his mind, he could see his route as if he were carrying a map. He went to Pinnacles National Monument, went up the Vazquez Monolith without an indecisive moment, and, on top, built a cairn around the brass tube. When Brower led a group to Shiprock in 1939, at least ten previous climbing parties had tried and failed there. Shiprock is a seven-thousand-foot monadnock that looks something like a schooner rising in isolation from the floor of the New Mexican desert. Brower studied photographs of Shiprock for many months, then planned an ornately complicated route—about three-quarters of the way up one side, then far down another side, then up a third and, he hoped, final side, to the top. That is how the climb went, without flaw, start to finish. Another brass tube. "I like mountains. I like granite. I particularly like the feel of the Sierra granite. When I climbed the Chamonix Aiguilles, the granite felt so much like the granite in the Yosemite that I felt right at home. Once, in the Sierra, when I was learning, I was going up the wall of a couloir and I put both hands and one knee on a rock. The rock moved, and fell. It crashed seventy-five feet below. One of my hands had shot upward, and with two fingers I caught a ledge. I pulled myself up, and I sat there on that ledge and thought for a long while. Why was I that stupid—to put that much faith in one rock? I have an urge to get up on top. I like to get up there and see around. A three-hundred-and-sixty-degree view is a nice thing to have. I like to recognize where I've been, and look for routes where I might go."
Encounters with the Archdruid

HOLLYWOOD CANNOT RESIST THE tetons. If you have seen Western movies, you have seen the Tetons. They have appeared in the background of countless pictures, and must surely be the most tectonically active mountains on film, drifting about, as they will, from Canada to Mexico, and from Kansas nearly to the coast. After the wagon trains leave Independence and begin to move westward, the Tetons soon appear on the distant horizon, predicting the beauty, threat, and promise of the quested land. After the wagons have been moving for a month, the Tetons are still out there ahead. Another fortnight and the Tetons are a little closer. The Teton Range is forty miles long and less than ten across—a surface area inverse in proportion not only to its extraordinary ubiquity but also to its grandeur. The Tetons—with Jackson Hole beneath them—are in a category with Mt. McKinley, Monument Valley, and the Grand Canyon of the Colorado River as what conservation organizations and the Washington bureaucracy like to call a scenic climax.
Rising from the Plains

Teton Range, Grand Teton National Park, Wyoming.

WE LEFT I-80 THERE AND BUCKED THE southwest wind, crossing the surprisingly flat mountain-crest terrain on a pair of ruts in the pink granite, which had crystals the size of silver dollars. The view from that high wide surface took in a large piece of the front of the Rockies, with the Never Summer Mountains standing out clearly in Colorado, to the south, and, to the west, the bright peaks of the Snowy Range. The Snowy Range—rising white above a dark high forest—appeared to be on top of the Medicine Bow Mountains. Remarkable as it seemed, that was the case. At the ten-thousand-foot level, the bottom of the Snowy Range rests on the broad flat top of the Medicine Bows like a sloop on water, its sails flying upward another two thousand feet. In the Miocene, the high flat Medicine Bow surface at the base of the Snowy Range was the level of maximum fill. In the fifty miles between the Snowy Range and our position on top of the Laramie Range lay the gulf of the excavated Laramie Plains. Our line of sight to the tree line of the Medicine Bows had been landscape in the Miocene. From twelve thousand feet it had gently sloped to about nine thousand where we were, and as we turned and faced east and gazed on down that mostly vanished plane we could all but see the Miocene surface continuing—as Love expressed it—"on out to East G-string."

Everywhere in the central Rockies, that highest level of basin fill touched the eminent ranges at altitudes that are now between ten and twelve thousand feet with results that are as beautiful as they are anomalous in the morphology of the world's mountains. In the Beartooths, for example, you can ascend a glacial valley that—in its U shape and high cirques— closely resembles any hanging valley in the Pennine Alps; but after you climb from ten to eleven to twelve thousand feet you do not find a Weisshorn fingering the sky. Instead, you move into an unexpectable physiographic setting, which, after steep slopes above a dry Wyoming basin, is lush and paradisal to the point of detachment from the world. Alpine meadows with meandering brooks are spread across a rolling but essentially horizontal scene, in part forested, in part punctuated with discrete stands of conifers and small cool lakes. The Medicine Bows are also like that—and the Uintas, the Bighorns. Their high flat surfaces, with peaks that seem to rest on them like crowns on tables, make no sense unless—as you look a hundred miles from one such surface to another across a deep dividing basin—you imagine earth instead of air: the Miocene fill, the continuous terrain. The high plateaus on the shoulders of the ranges, remaining from that broad erosional plane, have been given various names in the science, of which the most prominent at the moment is subsummit surface. "There's a plateau above Union Pass in the Wind River Range that's twelve thousand feet and flatter than a turd on a hot day," Love recalled, and went on to say that at such an altitude in flat country he sometimes becomes panicky—which does not happen if he is among craggy peaks, and seems to be a form of acrophobia directly related to the oddity of being in southern Iowa at twelve thousand feet.
Rising from the Plains

STANDING IN THAT WIND WAS LIKE standing in river rapids. It was a wind embellished with gusts, but, over all, it was primordially steady: a consistent

Stillwater Fork, Uinta National Forest, Utah.

Hoodos near
Farson, Wyoming,
with Wind River
Range in the
background.

southwest wind, which had been blowing that way not just through human history but in every age since the creation of the mountains—a record written clearly in wind-scored rock. Trees were widely scattered up there and, where they existed, appeared to be rooted in the rock itself. Their crowns looked like umbrellas that had been turned inside out and were streaming off the trunks downwind. "Wind erosion has tremendous significance in this part of the Rocky Mountain region," Love said. "Even down in Laramie, the trees are tilted. Old-timers used to say that a Wyoming wind gauge was an anvil on a length of chain. When the land was surveyed, the surveyors couldn't keep their tripods steady. They

had to work by night or near sunrise. People went insane because of the wind." His mother, in her 1905 journal, said that Old Hanley, passing by the Twin Creek school, would disrupt lessons by making some excuse to step inside and light his pipe. She also described a man who was evidently losing to the wind his struggle to build a cabin:

He was putting up a ridgepole when the wind was blowing. He looked up and saw the chipmunks blowing over his head. By and by, along came some sheep, dead. At last one was flying over who was not quite gone. He turned around and said, "Baa"— and then he was in Montana.

Erosion, giving the landscape its appearance, is said to be the work of water, ice, and wind; but wind is, almost everywhere, a minimal or negligible factor, with exceptional exceptions like Wyoming. Looking back across the interstate—north up the crest of the range—among ponderosas, aspens, and limber pines we could see the granites of Vedauwoo Glen, which had weathered out in large blocks, as granite does, along intersecting planes of weakness, while windborne grit had rounded off the corners of the blocks. Where some had tumbled and become freestanding, grit flying close above the ground had abraded them so rigorously that the subsummit surface was, in that place, a flat of giant mushrooms. The cliffs behind them also looked organic—high piles of rounded blocks, topped in many places by narrowly balanced boulders that were undercut almost to the point of falling. Love, contemplative, appeared to be puzzling out some deep question in geomorphology. At length, he said, "When wild horses defecate, they back up to a place where other wild horses have defecated, and so on, until they build turd towers, like those, in the air. Domestic horses do not do this."

At the Wyoming Information Center, beside Interstate 80 just south of Cheyenne, eleven picnic tables are enclosed in brick silos, and each silo has a picture window, so that visitors to Wyoming can picnic more or less al fresco and not be blown home. On the range, virtually every house has a shelter belt of trees—and for the most part the houses are of one story. Used tires cover the tops of mobile homes. Otherwise, wind tears off the roofs. Mary Kraus, a sedimentologist from the University of Colorado, got out of her car one day in north-central Wyoming and went to work on an outcrop. The wind

blew the car off a cliff. A propeller-drawn airplane that serves Wyoming is known as the Vomit Comet. When people step off it, they look like spotted slate.

"Most people today don't realize the power of wind and sand," Love said. "Roads are paved. But in the first fifty years of the Lincoln Highway you didn't like to travel west in the afternoon. You'd lose the finish on your car. Your windshield became so pitted you could hardly see out." The Highway Department has not yet paved the wind. On I-80, wind will capsize tractor-trailers. When snow

Balanced Rock of entrada sandstone, Arches National Park, Utah.

falls on Wyoming, its travels are only beginning. Snow snows again, from the ground up, moves along the surface in ground blizzards that can blind whole counties. Ground blizzards bury houses. In roadcuts, they make drifts fifty feet deep. The wind may return ahead of the plows and take the snow away. The old-timers used to say, "Snow doesn't melt here; it just wears out." Interstate 80 has been closed by snow in Wyoming in every month but July and August — sometimes closed for days. It is known as the Snow Chi Minh Trail. Before Amtrak dropped its Wyoming passenger service, people stranded on the Snow Chi Minh Trail used to abandon their cars and make their escape by train. The most inclement stretch of 80 is east of Rawlins where it skirts the tip of the Medicine Bows, where anemometers set on guardrails beside the highway frequently catch the wind exceeding the speed limit.

Rising from the Plains

MOVING ON WEST, ANOTHER DAY, WE crossed the Laramie Plains on I-80 through a world of what to me were surprising lakes. They were not glacial lakes or man-made lakes or — as in Florida — sinkhole lakes filling bowls of dissolved limestone. For the most part, they had no outlets, and were therefore bitter lakes — some alkaline, some saline, some altogether dry. Of Knadler Lake, about a mile long, Love said, "That's bitter water — sodium sulphate. It would physic you something awful." A herd of twenty antelopes galloped up the shore of Knadler Lake. Most of the lakes of the world are the resting places of rivers, where rivers seek their way through landscapes that have been roughed up and otherwise left chaotic by moving ice. Ice had never

covered the Laramie Plains. What, then, had dug out these lakes?

Love's response to that question was "What do you suppose?"

We had seen — a mile or two away — a hole in the ground eleven miles long, four miles wide, and deeper than the Yellow Sea. There were some puddles in it, but it did not happen to intersect any kind of aquifer, and basically it was dry. With a talent for understatement, the people of Laramie call it the Big Hollow. Geologists call it a deflation basin, a wind-scoured basin, or — more succinctly — a blowout. The wind at the Big Hollow, after finding its way into some weak Cretaceous shales, had in short order dug out four million acre-feet and blown it all away. Wind not only makes such basins but maintains them — usually within frameworks of resistant rock. On the Laramie Plains, the resistant rock is heavy quartzite gravel — Precambrian pieces of the Snowy Range which were brought to the plains as the beds of Pleistocene rivers. Wet or dry, all the lakes we passed had been excavated by the wind. It was a bright cloudless morning with a spring breeze. Spheres of tumbleweed, tumbling east, came at us on the interstate at high speed, like gymfuls of bouncing basketballs dribbled by the dexterous wind. "It's a Russian thistle," Love said. "It's one of nature's marvels. As it tumbles, seeds are exploded out."

Across the green plains, the Medicine Bow Mountains and the Snowy Range stood high, sharp, and clear, each so unlike the other that they gave the impression of actually being two ranges: in the middle distance, the flat-crested Medicine Bows, dark with balsam, spruce, and pine; and, in the far high background, the white and

Tumbleweeds.

treeless Snowy Range. That the one was in fact directly on top of the other was a nomenclatural Tower of Babel that contained in its central paradox the narrative of the Rockies: the burial of the ranges, the subsequent uplifting of the entire region, the exhumation of the mountains. As if to emphasize all that, people had not only named this single mountain range as if it were two but also bestowed upon the highest summit of the Snowy Range the name Medicine Bow Peak. It was up there making its point, at twelve thousand thirteen feet.

Rising from the Plains

NEAR ARLINGTON, AN ANOMALOUS piece of landscape reached straight out from the mountains like a causeway heading north. It was capped with stream gravel, brought off the mountains by furious rivers rushing through the tundras of Pleistocene time. The gravel had resisted subsequent erosion, while lighter stuff was washed away on either side. Geologists call such things pediments, and Love remarked that the one before us was "the most striking pediment in this region." In my mind's eye I could see the big braided rivers coming off the Alaska Range, thickly spreading gravel a mile wide, perhaps to preserve beneath them the scenes of former worlds. Where I-80 cut through the Arlington pediment, the Pleistocene gravel rested on Eocene sandstones, on red and green claystones; and they in turn covered conglomerates that came from the mountains when the mountains were new. One could read upward from one world to another: the boulders falling from rising mountains, the quiet landscapes after the violence stopped—all preserved in a perplexing memento from the climate of an age of ice.

Rising from the Plains

ALONG THE NEBRASKA-WYOMING LINE, in the region of the forty-first parallel, is a long lumpy break in the plains, called Pine Bluffs. It is rock of about the same age and story as Scotts Bluff, which is not far away. David Love—standing on top of Pine Bluffs—remarked that for a great many emigrants with their wagons and carts these had been the first breaks in the horizon west of Missouri. From the top of the bluffs, the emigrants had their first view of the front ranges of the Rockies, and the mountains gave them hope and courage. For our part, looking west from the same place, we could not see very far across the spring wildflowers into the swirling snow. The Laramie Range was directly ahead and the Never Summer Mountains off somewhere to the southwest—at ten and nearly thirteen thousand feet indeed a stirring sight, but not today. Love said a spring snowstorm was "sort of like a kiss—it's temporary, and it will go away." (That one stopped us for three days.)

Rising from the Plains

IN THE WESTERN OUTSKIRTS OF Rawlins, David Love pulled over onto the shoulder of the interstate, the better to fix the scene, although his purpose in doing so was not at all apparent. Rawlins reposed among low hills and prairie flats, and nothing in its setting would ever lift the stock of Eastman Kodak. In those western outskirts, we may have been scarcely a mile from the county courthouse, but we were very much back on the range—a dispassionate world of bare rock, brown grass, drab green patches of greasewood, and scattered colonies of sage. The interstate had lithified in 1965 as white concrete but was now dark with the remains of ocean algae, cremated and sprayed on the road. To the south were badlands—gullies and gulches, erosional debris. To the north were some ridgelines that ended sharply, like breaking waves, but the Rawlins Uplift had miserably fallen short in its bid to be counted among the Rocky Mountains. So why was David Love, who had the geologic map of Wyoming in his head, stopping here?

The rock that outcropped around Rawlins, he said, contained a greater spread of time than any other suite of exposed rocks along Interstate 80 between New York and San Francisco. We were looking at many moments in well over half the existence of the earth, and we were seeing—as it happened—a good deal more time than one sees in the walls of the Grand Canyon, where the clock stops at the rimrock, aged two hundred and fifty million years. The rock before us here at Rawlins reached back into the Archean eon and up to the Miocene epoch. Any spendthrift with a camera could aim it into that scene and—in a two-hundred-and-fiftieth of a second at f/16—capture twenty-six hundred million years. The most arresting thing in the picture, however, would be Rawlins' municipal standpipe—that white, squat water-storage tank over there on the hill.

The hill, though, was Archean granite and Cambrian sandstone and Mississippian limestone. If you could have taken pictures when *they* were forming, the collection would be something to see. There would be a deep and uncontinented ocean sluggish with amorphous scums (above cooling invisible magmas). There would be a risen continent reaching its coast, with rivers running over bare rock past not so much as a lichen. There would be rich-red soil on a broad lowland plain resembling Alabama (but near the equator). There would be clear, warm shelf seas.

There would also be a picture of dry hot dunes, all of them facing the morning sun—the rising Miocene sun. Other—and much older—dunes would settle a great question, for it is impossible to tell now whether they were just under or just out of water. They covered all of Wyoming and a great deal more, and may have been very much like the Empty Quarter of Arabia: the Tensleep-Casper-Fountain Pennsylvanian sands. There would be a picture, too, of a meandering stream, with overbank deposits, natural levees, cycads growing by the stream. Footprints the size of washtubs. A head above the trees. In the background, swamp tussocks by the shore of an oxbow lake. What was left of that picture was the Morrison formation—the Jurassic landscape of particularly dramatic dinosaurs—outcropping just up the road. There would be various views of the great Cretaceous seaway, with its plesiosaurs, its giant turtles, its crocodiles. There would be a picture from the Paleocene of a humid subtropical swamp, and a picture from the Eocene of gravel bars in a fast river running off a mountain onto lush subtropical plains, where puppy-size horses were hiding for their lives.

Such pictures, made in this place, could form a tall stack—scene after scene, no two of them alike. Taken together, of course—set one above another, in order—they would be the rock column for this part of Wyoming. They would correlate with what one would see in the well log of a deep-drilling rig. There would be hiatuses, to be sure. In the rock column, anywhere, more time is missing than is there; so much has been eroded away. Besides, the rock in the column is more apt to commemorate a moment—an eruption, a flood, a fallen drop of rain—than it is to report a millennium. Like a news broadcast, it is more often a montage of disasters than a cumulative record of time.

Rising from the Plains

NOW, AS WE CROSSED THE NORTH Platte River and ran on toward Rawlins in May, over the road were veils of blowing snow. This was Wyoming, not some nice mild place like Baffin Island — Wyoming, a landlocked Spitsbergen — and gently, almost imperceptibly, we were climbing. The snow did not obscure the structure. We were running above — and, in the roadcuts, among — strata that were leaning toward us, strata that were influenced by the Rawlins Uplift, which could be regarded as a failed mountain range. The Medicine Bow Mountains and the Sierra Madre stood off to the south, and while they and other ranges were rising this one had tried, too, but had succeeded merely in warping the flat land. The tilt of the strata was steeper than the road. Therefore, as we moved from cut to cut we were descending in time, down-section, each successive layer stratigraphically lower and older than the one before. Had this been a May morning a hundred million years ago, in Cretaceous time, we would have been many fathoms underwater, in a broad arm of the sea, which covered the continental platform — reached across the North American craton, the Stable Interior Craton — from the Gulf of Mexico to the Arctic Ocean. The North Platte, scratching out the present landscape, had worked itself down into some dark shales that had been black muds in the organic richness of that epicratonic sea. The salt water rose and fell, spread and receded through time — in Love's words, "advanced westward and then retreated, then advanced and retreated over and over again, leaving thick sequences of intertonguing sandstone and shale" — repeatedly

exposing fresh coastal plains, and as surely flooding them once more. In what has become dry mountain country, vegetation rioted in coastal swamps. They would have been like the Florida Everglades, the peat fens of East Anglia, or borders of the Java Sea, which stand just as temporarily, and after they are flooded by a rising ocean may be buried under sand and mud, and reported to the future as coal. There were seams of coal in the roadcuts, under the layers of sandstone and shale. The Cretaceous swamps were particularly abundant in this part of Wyoming. A hundred million years later, the Union Pacific Railroad would choose this right-of-way so it could fuel itself with the coal.

Rising from the Plains

LESS THAN A MILE UP THE ROAD, WE stopped again — at a low, flaky roadcut of Mowry shale. Progressing thus across Wyoming with David Love struck me as being analogous to walking up and down outside a theatre in the company of David Garrick. The classic plays — Teton, Beartooth, Wind River — were not out here on the street, but meanwhile these roadcuts were like posters, advertising the dramatic events, suggesting their narratives, fabrics, and structures. This Mowry shale had been organic mud of the Cretaceous seafloor, wherein the oil of the Frontier could have formed. It was a shale so black it all but smelled of low tide. In it, like mica, were millions of fish scales. It was interlayered with bentonite, which is a rock so soft it is actually plastic — pliable and porous, color of cream, sometimes the color of chocolate. Bentonite is volcanic tuff — decomposed, devitrified. So much volcanic debris has settled on Wyoming that bentonite is widespread and, in many places, more than ten feet thick. To some extent, it covers every basin. Also known as mineral soap, it has the bizarre ability to adsorb water up to fifteen times its own volume, and when this happens it offers to a tire about as much resistance as soft butter. Wet, swollen bentonite soil is known as gumbo. We were crossing badlands of the Bighorn Basin one time when a light shower fell, and the surface of the road changed in moments from dust to colloidal suspension. The wheels began to skid as if they were climbing ice. Four-wheel drive was no help. Many a geologist has walked out forty miles from a vehicle shipwrecked in gumbo. Bentonite is mined in Wyoming and sold to the rest of the world. Blessed is the land that can sell its mud. Bentonite is used in adhesives, automobile polish, detergent, and paint. It is in the drilling "mud" of oil rigs, sent down the pipe and through apertures in the bit to carry rock chips to the surface. It sticks to the walls of the drill hole and keeps out unwanted water. It is used to line irrigation ditches and reservoirs, and in facial makeup. Indians drove buffalo into swamps full of bentonite. It is an ingredient of insecticides, insect repellents, and toothpaste. It is used to clarify beer.

Rising from the Plains

IT WAS AT SOME MOMENT IN THE Pleistocene that humanity crossed what the geologist-theologian Pierre Teilhard de Chardin called the Threshold of Reflection, when something in people "turned back on itself and so to speak took an infinite leap forward. Outwardly, almost nothing in the organs had changed. But in depth, a great revolution had taken place: consciousness was now leaping and boiling in a space of super-sensory relationships and representations; and simultaneously consciousness was capable of perceiving itself in the concentrated simplicity of its faculties. And all this hap-

pened for the first time." Friars of another sort — evangelists of the so-called Environmental Movement — have often made use of the geologic time scale to place in perspective that great "leap forward" and to suggest what our reflective capacities may have meant to Mother Earth. David Brower, for example, the founder of Friends of the Earth and emeritus hero of the Sierra Club, has tirelessly travelled the United States for thirty years delivering what he himself refers to as "the sermon," and sooner or later in every talk he invites his listeners to consider the six days of Genesis as a figure of speech for what has in fact been four and a half billion years. In this adjustment, a day equals something like seven hundred and fifty million years, and thus "all day Monday and until Tuesday noon creation was busy getting the earth going." Life began Tuesday noon, and "the beautiful, organic wholeness of it" developed over the next four days. "At 4 P.M. Saturday, the big reptiles came on. Five hours later, when the redwoods appeared, there were no more big reptiles. At three minutes before midnight, man appeared. At one-fourth of a second before midnight, Christ arrived. At one-fortieth of a second before midnight, the Industrial Revolution began. We are surrounded with people who think that what we have been doing for that one-fortieth of a second can go on indefinitely. They are considered normal, but they are stark raving mad." Brower holds up a photograph of the world — blue, green, and swirling white. "This is the sudden insight from Apollo," he says. "There it is. That's all. We see through the eyes of the astronauts how fragile our life really is." Brower has computed that we are driving through the earth's resources at a rate comparable to a man's driving an automobile a hundred

and twenty-eight miles an hour—and he says that we are accelerating.

In like manner, geologists will sometimes use the calendar year as a unit to represent the time scale, and in such terms the Precambrian runs from New Year's Day until well after Halloween. Dinosaurs appear in the middle of December and are gone the day after Christmas. The last ice sheet melts on December 31st at one minute before midnight, and the Roman Empire lasts five seconds. With your arms spread wide again to represent all time on earth, look at one hand with its line of life. The Cambrian begins in the wrist, and the Permian Extinction is at the outer end of the palm. All of the Cenozoic is in a fingerprint, and in a single stroke with a medium-grained nail file you could eradicate human history. Geologists live with the geologic scale. Individually, they may or may not be alarmed by the rate of exploitation of the things they discover, but, like the environmentalists, they use these repetitive analogies to place the human record in perspective—to see the Age of Reflection, the last few thousand years, as a small bright sparkle at the end of time. They often liken humanity's presence on earth to a brief visitation from elsewhere in space, its luminous, explosive characteristics consisting not merely of the burst of population in the twentieth century but of the whole millennial moment of people on earth—a single detonation, resembling nothing so much as a nuclear implosion with its successive neutron generations, whole generations following one another once every hundred-millionth of a second, temperatures building up into the millions of

degrees and stripping atoms until bare nuclei are wandering in electron seas, pressures building up to a hundred million atmospheres, the core expanding at five million miles an hour, expanding in a way that is quite different from all else in the universe, unless there are others who also make bombs.

The human consciousness may have begun to leap and boil some sunny day in the Pleistocene, but the race by and large has retained the essence of its animal sense of time. People think in five generations—two ahead, two behind—with heavy concentration on the one in the middle. Possibly that is tragic, and possibly there is no choice. The human mind may not have evolved enough to be able to comprehend deep time. It may only be able to measure it. At least, that is what geologists wonder sometimes, and they have imparted the questions to me. They wonder to what extent they truly sense the passage of millions of years. They wonder to what extent it is possible to absorb a set of facts and move with them, in a sensory manner, beyond the recording intellect and into the abyssal eons. Primordial inhibition may stand in the way. On the geologic time scale, a human lifetime is reduced to a brevity that is too inhibiting to think about. The mind blocks the information. Geologists, dealing always with deep time, find that it seeps into their beings and affects them in various ways. They see the unbelievable swiftness with which one evolving species on the earth has learned to reach into the dirt of some tropical island and fling 747s into the sky. They see the thin band in which are the all but indiscernible stratifications of Cro-Magnon, Moses, Leonardo, and now. Seeing a race unaware of its own instantaneousness in time, they can reel off all the species that have come and gone, with emphasis on those that have specialized themselves to death.

In geologists' own lives, the least effect of time is that they think in two languages, function on two different scales.

"You care less about civilization. Half of me gets upset with civilization. The other half does not get upset. I shrug and think, So let the cockroaches take over."

"Mammalian species last, typically, two million years. We've about used up ours. Every time Leakey finds something older, I say, 'Oh! We're overdue.' We will be handing the dominant-species-on-earth position to some other group. We'll have to be clever not to."

"A sense of geologic time is the most important thing to suggest to the non-geologist: the slow rate of geologic processes, centimetres per year, with huge effects, if continued for enough years."

"A million years is a short time—the shortest worth messing with for most problems. You begin tuning your mind to a time scale that is the planet's time scale. For me, it is almost unconscious now and is a kind of companionship with the earth."

"It didn't take very long for those mountains to come up, to be deroofed, and to be thrust eastward. Then the motion stopped. That happened in maybe ten million years, and to a geologist that's really fast."

"If you free yourself from the conventional reaction to a quantity like a million years, you free yourself a bit from the boundaries of human time. And then in a way you do not live at all, but in another way you live forever."

Basin and Range

LAKES ARE SO EPHEMERAL THAT THEY are seldom developed in the geologic record. They are places where rivers bulge, as a temporary consequence of topography. Lakes fill in, drain themselves, or just evaporate and disappear. They don't last. The Great Lakes are less than twenty thousand years old. The Great Salt Lake is less than twenty thousand years old. When Lake Gosiute took in the finishing touch of sediment that ended its life, it was eight million years old.

West of Rock Springs, we came to an escarpment known as White Mountain, standing a thousand feet above the valley of Killpecker Creek. In no tectonic sense was this a true mountain—a folded-and-faulted, volcanic, or overthrust mountain. This was just a Catskill, a Pocono, a water-sliced segment of layered flat rock, a geological piece of cake. In fact, it was the bed of Lake Gosiute, and contained almost all of the eight million years. Apparently, the initial freshwater lake eventually shrank, became bitter and saline, and intermittently may have gone dry. Later, as the climate remoistened, water again filled the basin, and the lake reached its greatest size. As we looked at White Mountain, we could see these phases. It was the dry, salt-lake interval in the middle—straw and hay pastels so pale they were nearly white—that had given the bluff its name. The streams that had opened it to view were lying at its base. Killpecker Creek (full of saltpeter) flowed into Bitter Creek, and that soon joined the Green River.

Down the road a couple of miles was a pair of tunnels—snake eyes in the lakebed. They were one of only three sets of tunnels on Interstate 80 between New York and San Francisco, but they had to be there in the nose of White Mountain, or the interstate, flexing left, would destroy the town of Green River. Tower sandstone stood on the ridgeline in castellated buttes. With each mile, they

increased in number, like buildings on the outskirts of a city. Off to the left was the island from which the geologist John Wesley Powell—seven years before the battle of the Little Bighorn—set off in a flotilla of dinghies to follow the Green River into its master stream, and to survive some of the preeminent rapids of North America on the first known voyage through the Grand Canyon. A huge sandstone broch stood in brown shale above the tunnels, which penetrated the lakebed's saline phase. If the Great Salt Lake, which has been freshening in recent years, should follow the biography of Gosiute, it will swell up to the size of Huron and, as it once did, spill to the north over Red Rock Pass and pour down the Snake River Plain. Inundating much of Utah and some of Nevada, it will send the Mormons back to New York, saying, "That was the place."
Rising from the Plains

BASIN. FAULT. RANGE. BASIN. FAULT. Range. A mile of relief between basin and range. Stillwater Range. Pleasant Valley. Tobin Range. Jersey Valley. Sonoma Range. Pumpernickel Valley. Shoshone Range. Reese River Valley. Pequop Mountains. Steptoe Valley. Ondographic rhythms of the Basin and Range. We are maybe forty miles off the interstate, in the Pleasant Valley basin, looking up at the Tobin Range. At the nine-thousand-foot level, there is a stratum of cloud against the shoulders of the mountains, hanging like a ring of Saturn. The summit of Mt. Tobin stands clear, above the cloud. When we crossed the range, we came through a ranch on the ridgeline where sheep were fenced around a running brook and bales of hay were bright green. Junipers in the mountains were thickly hung with berries, and the air was unadulterated gin.

Rocks near Green River, Wyoming.

This country from afar is synopsized and dismissed as "desert"—the home of the coyote and the pocket mouse, the side-blotched lizard and the vagrant shrew, the MX rocket and the pallid bat. There are minks and river otters in the Basin and Range. There are deer and antelope, porcupines and cougars, pelicans, cormorants, and common loons. There are Bonaparte's gulls and marbled godwits, American coots and Virginia rails. Pheasants. Grouse. Sandhill cranes. Ferruginous hawks and flammulated owls. Snow geese. This Nevada terrain is not corrugated, like the folded Appalachians, like a tubal air mattress, like a rippled potato chip. This is not—in that compressive manner—a ridge-and-valley situation. Each range here is like a warship standing on its own, and the Great Basin is an ocean of loose sediment with these mountain ranges standing in it as if they were members of a fleet without precedent, assembled at Guam to assault Japan. Some of the ranges are forty miles long, others a hundred, a hundred and fifty. They point generally north. The basins that separate them—ten and fifteen miles wide—will run on for fifty, a hundred, two hundred and fifty miles with lone, daisy-petalled windmills standing over sage and wild rye. Animals tend to be content with their home ranges and not to venture out across the big dry valleys. "Imagine a chipmunk hiking across one of these basins," Deffeyes remarks. "The faunas in the high ranges here are quite distinct from one to another. Animals are isolated like Darwin's finches in the Galápagos. These ranges are truly islands."

Supreme over all is silence. Discounting the cry of the occasional bird, the wailing of a pack of coyotes, silence—a great spatial silence—is pure in the Basin and Range. It is a soundless immensity with mountains in it. You stand, as we do now, and look up at a high mountain front, and turn your head and look fifty miles down the valley, and there is utter silence. It is the silence of the winter forests of the Yukon, here carried high to the ridgelines of the ranges. As the physicist Freeman Dyson has written in *Disturbing the Universe,* "It is a soul-shattering silence. You hold your breath and hear absolutely nothing. No rustling of leaves in the wind, no rumbling of distant traffic, no chatter of birds or insects or children. You are alone with God in that silence. There in the white flat silence I began for the first

time to feel a slight sense of shame for what we were proposing to do. Did we really intend to invade this silence with our trucks and bulldozers and after a few years leave it a radioactive junkyard?"

AS WE WOUND DOWN THE MOUNTAIN at the end of the day, we stopped to regard the silent valley—the seventy miles of basin under a rouge sky, the circumvallate mountains, and, the better part of a hundred miles away, Sonoma Peak, of the Sonoma Range. Deffeyes said, "If you reduced the earth to the size of a baseball, you couldn't feel that mountain. With a telephoto lens, you could convince someone it was Everest." Even at this altitude, the air was scented powerfully with sage. There was coyote scat at our feet. In the dark, we drove back the way we had come, over the painted cattle guards and past jackrabbits dancing in the road, pitch-dark, and suddenly a Black Angus was there, standing broadside, middle of the road. With a scream of brakes, we stopped. The animal stood still, thinking, its eyes unmoving—a wall of beef. We moved slowly after that, and even more slowly when a white sphere materialized on our right in the moonless sky. It expanded some, like a cloud. Its light became so bright that we stopped finally and got out and looked up in awe. A smaller object, also spherical, moved out from within the large one, possibly from behind it. There was a Saturn-like ring around the smaller sphere. It moved here and there beside the large one for a few minutes and then went back inside. The story would be all over the papers the following day. The *Nevada State Journal* would describe a "Mysterious Ball of Light" that had been reported by various people at least a hundred miles in every

White granite in
Bells Canyon,
Lone Peak
Wilderness, Utah.

direction from the place where we had been. "By this time we decided to get the hell out of there," a couple of hunters reported, "and hopped in our pickup and took off. As we looked back at it, we saw a smaller craft come out of the right lower corner. This smaller craft had a dome in the middle of it and two wings on either side, but the whole thing was oval-shaped." Someone else had said, "I thought it was an optical illusion at first, but it just kept coming closer and closer so that I could see it wasn't an illusion. Then something started coming out of the side of it. It looked like a star, and then a ring formed around it. A kind of ring like you'd see around Saturn. It didn't make any noises, and then it vanished."

"Now we're both believers," said one of the hunters. "And I don't ever want to see another one. We're pretty good-sized men and ain't scared of nothing except for snakes and now flying saucers."

After the small sphere disappeared, the large one rapidly faded and also disappeared. Deffeyes and I were left on the roadside among the starlighted eyes of dark and motionless cattle. "Copernicus took the world out of the center of the universe," he said. "Hutton took us out of a special place somewhere near the beginning of things and left us awash in the middle of the immensity of time. An extraterrestrial civilization could show us where we are with regard to the creation of life."

Basin and Range

5. NOBLE METALS

EFFEYES' PURPOSES IN COMING to Nevada are pure and noble. His considerable energies appear to be about equally divided between the pursuit of pure science and the pursuit of noble metal. In order to enloft mankind's understanding of the basins, he has been taking paleomagnetic samples of basin sediments. He seeks insight into the way in which the rifting earth comes apart. He wants to perceive the subtle differences in the histories of one fault block and another. His ideas about silver, on the other hand, may send his children to college. This is, after all, Nevada, whose geology bought the tickets for the Spanish-American War. George Hearst found his fortune in the ground here. There were silver ores of such concentration that certain miners did nothing more to the heavy gray rocks than pack them up and ship them to Europe. To be sure, those days and those rocks are gone, but it has crossed the mind of Deffeyes that there may be something left for Deffeyes. Banqueting Sybarites surely did not lick their plates.

Basin and Range

FOR EIGHT MILLION YEARS, DEFFEYES was saying, as the crustal blocks inexorably pulled apart here and springs boiled up along the faults, silver had been deposited throughout the Basin and Range. The continually growing mountains sometimes fractured their own ore deposits, greatly complicating the sequence of events and confusing the picture for anyone who might come prospecting for ores. There was another phenomenon, however, that had once made prospecting dead simple. Erosion, breaking into hot-spring and vein deposits, concentrated the silver. Rainwater converted silver sulphides to silver chloride, heavy stuff that stayed right where it was and — through thousands of millennia — increased in concentration as more rain fell. Such deposits, richer than an Aztec dream, were known to geologists as supergene enrichments. Miners called them surface bonanzas. In the eighteen-sixties, and particularly in the eighteen-seventies, they were discovered

Virgins Bower grows on mine tailings in Ophir, Utah.

Ward Charcoal
Ovens, Ward ghost
town, Nevada.

in range after range. A big supergene enrichment might be tens of yards wide and a mile long, lying at or near the surface. Instant cities appeared beside them, with false-front saloons and tent ghettos, houses of sod, shanties made of barrels. The records of these communities suggest uneven success in the settling of disputes between partners over claims: "Davison shot Butler through the left elbow, breaking the bone, and in turn had one of his toes cut off with an axe." They were places with names like Hardscrabble, Gouge Eye, Battle Mountain, Treasure Hill. By the eighteen-nineties, the boom was largely over and gone. During those thirty years, there were more communities in Nevada than there are now. "Silver is our most depleted resource, because it gave

itself away," said Deffeyes, looking mournful. "You didn't need a Ph.D. in geology to find a supergene enrichment."

All you needed was Silver Jim. Silver Jim was a Paiute, and he, or a facsimile, took you up some valley or range and showed you grayish rock with touches of green that had a dull waxy lustre like the shine on the horn of a cow. Horn silver. It was just lying there, difficult to lift. Silver Jim could show you horn silver worth twenty-seven thousand dollars a ton. Those were eighteen-sixties dollars and an uninflatable ton. You could fill a wheelbarrow and go down the hill with five thousand dollars' worth of silver. Three or four years ago, a miner friend of Deffeyes who lives in Tombstone, Arizona, happened to find on his own property an overlooked fragment of a supergene enrichment, a narrow band no more than a few inches thick, six feet below the cactus. Knocking off some volcanic overburden with a front-end loader, the miner went after this nineteenth-century antique and fondly dug it out by hand. He said to his children, "Pay attention to what I'm doing here. Look closely at the rock. We will never see this stuff again." In a couple of hours of a weekend afternoon, he took twenty thousand dollars from the ground.

We were off on dirt roads now with a cone of dust behind us, which Deffeyes characterized as the local doorbell. He preferred not to ring it. This talkative and generous professor—who ordinarily shares his ideas as rapidly as they come to him, spilling them out in bunches like grapes—was narrow-eyed with secrecy today. He had stopped at a courthouse briefly, and—an antic figure, with his bagging sweater and his Beethoven hair— had revealed three digits to a county clerk in requesting to see a registry of claims. The claims were coded in six digits. Deffeyes kept the fourth, fifth, and sixth to himself like cards face down on a table. He found what he sought in the book of claims. Now, fifty miles up the valley, we had long since left behind us its only town, with its Odd Fellows Hall, its mercantile company, its cottonwoods and Lombardy poplars; and there were no houses, no structures, no cones of dust anywhere around us. The valley was narrowing. It ended where ranges joined. Some thousands of feet up the high face of a distant and treeless mountain we saw an unnaturally level line.

"Is that a road?" I asked him.

"That's where we're going," he said, and I wished he hadn't told me.

Looking up there, I took comfort in the reflection that I would scarcely be the first journalist to crawl out on a ledge in the hope of seeing someone else get rich. In 1869, the editor of the New York *Herald,* looking over his pool of available reporters, must have had no difficulty in choosing Tom Cash to report on supergene enrichments. Cash roved Nevada. He reported from one place that he took out his pocketknife and cut into the wall of a shaft, removing an ore of such obviously high assay that he could roll it in his fingers and it would not crumble. Cash told the mine owner that he feared being accused of exaggeration—"of making false statements, puffing"—with resulting damage to his journalistic reputation. There was a way to avoid this, he confided to the miner. "I would like to take a sample with me of some of the richest portions." The miner handed him a fourteen-pound rock containing about a hundred and fifty troy ounces of silver (seventy-three per cent). In the same year, Albert S. Evans, writing in the San Francisco *Alta California,* described a visit with a couple of bankers and a geologist to a claim in Nevada where he was lowered on a rope into a mine. "The light of our candles disclosed great black sparkling masses of silver on every side. The walls were of silver, the roof over our heads was of silver, and the very dust that filled our lungs and covered our boots and clothing with a gray coating was of fine silver. We were told that in this chamber a million dollars worth of silver lies exposed to the naked eye and our observations confirm the statement. How much lies back of it, Heaven only knows."

Heaven knew exactly. For while the supergene enrichments—in their prodigal dispersal through the Basin and Range— were some of the richest silver deposits ever discovered in the world, they were

also the shallowest. There was just so much lying there, and it was truly bonanzan—to print money would take more time than to pick up this silver—but when it was gone it was gone, and it went quickly. Sometimes—as in the Comstock Lode in Virginia City—there were "true veins" in fissures below, containing silver of considerable value if more modest assay, but more often than not there was nothing below the enrichment. Mining and milling towns developed and died in less than a decade.
Basin and Range

WE TURNED A LAST CORNER, WITH OUR inner wheels resting firmly on the road and the two others supported by Deffeyes' expectations. Now we were moving along one wall of a big V-shaped canyon that eventually became a gulch, a draw, a crease in the country, under cottonwoods. In the upper canyon, some hundreds of acres of very steep mountainside were filled with holes and shafts, hand-forged ore buckets, and old dry timbers. There were square nails in the timbers. An ore bucket was filled with square nails. "Good litter," Deffeyes said, and we walked uphill past the mine and along a small stream into the cottonwoods. The stream was nearly dry. Under the cottonwoods were the outlines of cabins almost a century gone. Here at seven thousand feet in this narrow mountain draw had lived a hundred people, who had held their last election a hundred years ago. They had a restaurant, a brewery, a bookstore. They had seven saloons. And now there was not so much as one dilapidated structure. There were only the old unhappy cottonwoods, looking alien and discontented over the moist bed of the creek. Sixteen stood there, twisted, surviving—most of them over four feet thick. "Those cottonwoods

try an environmentalist's soul," Deffeyes said. "They transpire water like running fountains. If you were to cut them down, the creek would run. Cottonwoods drink the Humboldt. Some of the tension in this country is that miners need water. Getting rid of trees would preserve water. By the old brine-and-mercury method, it took three tons of water to mill one ton of ore. There was nothing like that in this creek. They had to take the ore from here to a big enough stream, and that, as it happens, was a twelve-mile journey using mules. They would have gone out of here with only the very best ore. There was probably a supergene enrichment here over a pretty good set of veins. They took what they took and were gone in six years."
Basin and Range

MOST OF THE PASS WAS COVERED WITH snow, but there were some patches of bare ground, and these were blue, green, red, yellow, and white with wild flowers. The air felt and smelled like the first warm, thaw-bringing day in spring in Vermont, and, despite the calendar, spring was now the season at that altitude in the North Cascades, and summer and fall would come and go in the few weeks remaining before the first big snow of September. Brower had dropped his pack and was sitting on a small knoll among the flowers. Park and I and Brigham and Snow dropped our own packs, and felt the sudden coolness of air reaching the sweatlines where the packs had been—and the inebriate lightness that comes, after a long climb, when the backpack is suddenly gone. The ground Brower was sitting on was ten or fifteen feet higher than the ground on which we stood, and as we went up to join him our eyes at last moved above the ridgeline, and for the first time we could

see beyond it. What we saw made us all stop.

One of the medical students said, "Wow!"

I said slowly, the words just involuntarily falling out, "My God, look at that."

Across a deep gulf of air, and nearly a mile higher than the ground on which we stood, eleven miles away by line of sight, was Glacier Peak — palpable, immediate, immense. In the direction we were looking, we could see perhaps two hundred square miles of land, and the big mountain dominated that scene in the way that the Jungfrau dominates the Bernese Alps. Glacier Peak had originally been a great symmetrical cone, and that was still its basic shape, but it had been monumentally scarred, from within and without. It once exploded. Pieces of it landed in what is now Idaho, and other pieces landed in what is now Oregon. The ice sheet mauled it. Rivers from its own glaciers cut grooves in it. But it had remained, in sil-

Cottonwood along the bed of the Green River in Wyoming.

houette, a classic mountain, its lines sweeping up beyond its high shoulder — called Disappointment Peak — and converging acutely at the summit. The entire upper third of the mountain was white. And below the snow and ice, black-green virgin forest continued all the way down to the curving valley of the Suiattle River, a drop of eight thousand feet from the peak. Spread around the summit like huge, improbable petals were nine glaciers — the Cool Glacier, the Scimitar Glacier, the Dusty Glacier, the Chocolate Glacier — and from each of these a white line of water ran down through the timber and into the Suiattle. To our right, on the near side of the valley, another mountain — Plummer Mountain — rose up about two-thirds as high, and above its timberline its snowless faces of rock were, in the sunlight, as red as rust. Around and beyond Glacier Peak, the summits of other mountains, random and receding, led the eye away to the rough horizon and back to Glacier Peak.

Brower said, without emphasis, "That is what is known in my trade as a scenic climax."

Near the southern base of Plummer Mountain and in the deep valley between Plummer Mountain and Glacier Peak — that is, in the central foreground of the view that we were looking at from Cloudy Pass — was the lode of copper that Kennecott would mine, and to do so the company would make an open pit at least two thousand four hundred feet from rim to rim.

Park said, "A hole in the ground will not materially hurt this scenery."

Brower stood up. "None of the experts on scenic resources will agree with you," he said. "This is one of the few remaining great wildernesses in the lower forty-eight. Copper is not a transcendent value here."

"Without copper, we'd be in a pretty sorry situation."

"If that deposit didn't exist, we'd get by without it."

"I would prefer the mountain as it is, but the copper is there."

"If we're down to where we have to take copper from places this beautiful, we're down pretty far."

"Minerals are where you find them. The quantities are finite. It's criminal to waste minerals when the standard of living of your people depends upon them. A mine cannot move. It is fixed by nature. So it has to take precedence over any other use. If there were a copper deposit in Yellowstone Park, I'd recommend mining it. Proper use of minerals is essential. You have to go get them where they are. Our standard of living is based on this."

"For a fifty-year cycle, yes. But for the long term, no. We have to drop our standard of living, so that people a thousand years from now can have any standard of living at all."

A breeze coming off the nearby acres of snow felt cool but not chilling in the sunshine, and rumpled the white hair of the two men.

"I am not for penalizing people today for the sake of future generations," Park said.

"I really am," said Brower. "That's where we differ."

"Yes, that's where we disagree. In 1910, the Brazilian government said they were going to preserve the iron ore in Minas Gerais, because the earth would run short of it in the future. People — thousands and thousands of people in Minas Gerais — were actually starving, and they were living over one of the richest ore deposits in the world, a fifteen-billion-ton reserve. They're mining it now, and people there are prospering. But in the past it was poor consolation to people who were going hungry to say that in the future it was going to be better. You have to use these things when you have them. You have to know where they are, and use them. People, in the future, will go for the copper here."

"The kids who are in Congress in the future should make that decision, and if it's theirs to make I don't think they'll go for the copper here," Brower said.

"Sure they will. They'll have to, if people are going to expect to have telephones, electric lights, airplanes, television sets, radios, central heating, air-conditioning, automobiles. And you *know* people will want these things. I didn't invent them. I just know where the copper is."

Brower swung his pack up onto his back. "Pretend the copper deposit down there doesn't exist," he said. "Then what would you do? What are you going to do when it's gone?"

"You're trying to make everything wilderness," Park said.

"No, I'm not. I'm trying to keep at least two per cent of the terrain as wilderness."

"Two per cent is a lot."

"Two per cent is under pavement."

"Basically, our difference is that I feel we can't stop all this — we must direct it. You feel we must stop it."

"I feel we should go back, recycle, do things over again, and do better, even if it costs more. We mine things and don't use them again. We coat the surface of the earth — with beer cans and chemicals, asphalt and old television sets."

"We *are* recycling copper, but we don't have enough."

"When we knock buildings down, we don't take the copper out. Every building that comes down could be a copper mine. But we don't take the copper out. We go

after fresh metal. We destroy that mountain."

"How can you ruin a mountain like Glacier Peak?" Park lifted his pick toward the mountain. "You *can't* ruin it," he went on, waving the pick. "Look at the Swiss mountains. Who could ruin *them?* A mine would not hurt this country — not with proper housekeeping."

Brower started on down the trail. We retrieved our packs and caught up with him. About five hundred feet below us and a mile ahead was another pass — Suiattle Pass — and to reach it we had to go down into a big ravine and up the other side. There were long silences, measured by the sound of boots on the trail. From time to time, the pick rang out against a rock.

Brower said, "Would America have to go without much to leave its finest wilderness unspoiled?"

We traversed a couple of switchbacks and approached the bottom of the ravine. Then Park said, "Where they are more easily accessible, deposits have been found and are being — or have been — mined."

We had seen such a mine near Lake Chelan, in the eastern part of the mountains. The Howe Sound Mining Company established an underground copper mine there in 1938, built a village and called it Holden. The Holden mine was abandoned in 1957. We had hiked past its remains on our way to the wilderness area. Against a backdrop of snowy peaks, two flat-topped hills of earth detritus broke the landscape. One was the dump where all the rock had been put that was removed before the miners reached the ore body. The other consisted of tailings — crushed rock that had been through the Holden mill and had yielded copper. What remained of the mill itself was a macabre skeleton of bent, twisted, rusted beams.

Wooden buildings and sheds were rotting and gradually collapsing. The area was bestrewn with huge flakes of corrugated iron, rusted rails, rusted ore carts, old barrels. Although there was no way for an automobile to get to Holden except by barge up Lake Chelan and then on a dirt road to the village, we saw there a high pile of gutted and rusted automobiles, which themselves had originally been rock in the earth and, in the end, in Holden, were crumbling slowly back into the ground.

Park hit a ledge with the pick. We were moving up the other side of the ravine now. The going was steep, and the pace slowed. Brower said, "We saw that at Holden."

I counted twenty-two steps watching the backs of Brower's legs, above the red tops of gray socks. He was moving slower than I would have. I was close behind him. His legs, blue-veined, seemed less pink than they had the day before. They were sturdy but not athletically shapely. Brower used to put food caches in various places in the High Sierra and go from one to another for weeks at a time. He weighed two hundred and twelve pounds now, and he must have wished he were one-eighty.

Park said, "Holden is the sort of place that gave mining a bad name. This has been happening in the West for the past hundred years, but it doesn't have to happen. Poor housekeeping is poor housekeeping wherever you find it. I don't care if it's a mine or a kitchen. Traditionally, when mining companies finished in a place they just walked off. Responsible groups are not going to do that anymore. They're not going to leave trash; they're not going to deface the countryside. Think of that junk! If I had enough money, I'd come up here and clean it up."

I thought how neat Park's house, his lawn, and his gardens are—his roses, his lemon tree, his two hundred varieties of cactus. The name of the street he lives on is Arcadia Place. Park is a member of the Cactus and Succulent Society of America. He hit a fallen tree with the hammer end.

"It's one god-awful mess," Brower said.

"That old mill could be cleaned up," Park said. "Grass could be planted on the dump and the tailings."

Suiattle Pass was now less than a quarter mile ahead of us. I thought of Brower, as a child, on his first trip to the Sierra Nevada. His father drove him there from Berkeley in a 1916 Maxwell. On the western slopes, they saw both the aftermath and the actual operations of hydraulic mining for gold. Men with hoses eight inches in diameter directed water with such force against the hillsides that large parts of the hills themselves fell away as slurry.

"Holden was abandoned in 1957, and no plants of any kind have caught on the dump and the tailings," Brower said.

Holden, in its twenty years of metal production, brought out of the earth ten million tons of rock—enough to make a hundred thousand tons of copper, enough to wire Kansas City.

Park said, "You could put a little fertilizer on—something to get it started."

When we reached the pass, we stood for a moment and looked again at Glacier Peak and, far below us, the curving white line of the Suiattle. Park said, "When you create a mine, there are two things you can't avoid: a hole in the ground and a dump for waste rock. Those are two things you can't avoid."

Brower said, "Except by not doing it at all."

Encounters with the Archdruid

TWENTY MILES OUT OF WINNEMUCCA, and the interstate is dropping south toward the Humboldt Range. A coyote runs along beside the road. It is out of its element, tongue out, outclassed, under minimum speed. Deffeyes says that most ranges in the Basin and Range had one or two silver deposits in them, if any, but the Humboldts had five. We have also entered the bottomlands of the former Lake Lahontan. The hot-springs map shows more activity in this part of the province. Extension of the earth's crust has been

somewhat more pronounced here, Deffeyes explains, and hence there are more ore deposits. He feels that when a seaway opens up, the spreading center will be somewhere nearby. Or possibly back in Utah, in the bed of Lake Bonneville. "But this one has better connections."

"Connections?"

"Death Valley. Walker Lake. Carson Sink." An Exxon map of the western United States is spread open on the seat between us. He runs his finger from Death Valley to Carson Sink and on northward

Ruby Mountains of the Humboldt National Forest, Nevada.

**Ruby Crest in the
Humboldt Range.**

to cross the interstate at Lovelock. "The ocean will open here," he repeats. "Or in the Bonneville basin. I think here."

A few miles off the road is the site of a planned community dating from the nineteen-sixties. It was to have wide streets and a fountained square, but construction was delayed and then indefinitely postponed. Ghostless ghost town, it had been named Neptune City.

With the river on our right, we round the nose of the Humboldt Range, as did the Donner Party and roughly a hundred and sixty-five thousand other people, in a seventeen-year period, heading in their wagons toward Humboldt Sink, Carson Sink, and the terror of days without water. But first, as we do now, they came into broad green flats abundantly fertile with grass, knee-high grass, a fill for the oxen, the last gesture of the river before it vanished into the air. The emigrants called this place the Big Meadows of the Humboldt, and something like two hundred and fifty wagons would be resting here at any given time.

"There was a sea here in the Triassic," Deffeyes remarks. "At least until the Sonomia terrain came in and sutured on. The sea was full of pelagic squid, and was not abyssal, but it was deep enough so the bottom received no sunlight, and bottom life was not dominant."

"How do you know it was not dominant?"

"Because I have looked at the siltstones

and the ammonites in them, and that is what I see there."

Visions of oceans before and behind us in time, we roll on into Lovelock. SLOW—DUST HAZARD. Lovelock, Nevada 89419. There are cumulus snow clouds overhead and big bays of blue in the cold sky, with snow coming down in curtains over the Trinity Range, snow pluming upward over the valley like smoke from a runaway fire. Lovelock was a station of the Overland Stage. It became known throughout Nevada as "a good town with a bad water supply." An editor of the Lovelock *Review-Miner* wrote in 1915, "There is little use in trying to induce people to locate here until the water question is settled. . . . Maybe the water does not kill anyone, but it certainly drives people away." In 1917, Lovelock was incorporated as a third-class city, and one of its first acts was to enforce a ban on houses of prostitution within twelve hundred feet of the Methodist Episcopal Church. Another was a curfew. Another ordered all city lights turned off when there was enough moon.

JAX CASINO LIBERAL SLOTS

•

TWO STIFFS SELLING GAS AND MOTEL

•

WATER SUPPLY FROM PRIVATE WELL

•

LOVELOCK SEED COMPANY
GRAINS AND FEED

Here in the Big Meadows of the Humboldt, the principal employer is the co-op seed mill on the edge of town, which sends alfalfa all over the world.

On the sidewalks are men in Stetsons, men in three-piece suits, men in windbreakers, tall gaunt overalled men with beards. There are women in Stetsons, boots, and jeans. A thin young man climbs

out of a pickup that is painted in glossy swirls of yellow and purple, and has a roll bar, balloon tires, headphones, and seventeen lights.

There are terraces of Lake Lahontan above the ballfield of the Lovelock Mustangs. Cattle graze beside the field. The Ten Commandments are carved in a large piece of metamorphosed granite outside the county courthouse.

NO. 10:
THOU SHALT NOT COVET THY
NEIGHBOR'S WIFE,
NOR HIS MANSERVANT, NOR HIS
MAIDSERVANT,
NOR HIS CATTLE

•

BRAZEN ONAGER—BAR—BUD—PIZZA

•

WHOO-O-A MOTEL

"Lovelock was a person's name," Deffeyes cautions.

LOVELOCK MERCANTILE

The name is fading on the cornice of Lovelock Mercantile. It was built in 1905, expanded in 1907, is the bus stop now, liquor store, clothing store, grocery store, real-estate office, bakery, Western Union office—all in one room. There is a sign on one of the columns that hold up the room:

WE CANNOT ACCEPT
GOVERNMENT MEAL TICKETS

Across the valley is a huge whitewash "L" on a rock above the fault scar of the Humboldt Range.

We go into Sturgeon's Log Cabin restaurant and sit down for coffee against a backdrop of rolling cherries, watermelons, and bells. A mountain lion in a glass case. Six feet to the tip of the tail. Shot by Daniel (Bill) Milich, in the Tobin Range.

I hand Deffeyes the Exxon map and ask him to sketch in for me the opening of the new seaway, the spreading center as he sees it coming. "Of course, all the valleys in the Great Basin are to a greater or lesser extent competing," he says. "But I'd put it where I said—right here." With a pencil he begins to rough in a double line, a swath, about fifteen miles wide. He sketches it through the axis of Death Valley and up into Nevada, and then north by northwest through Basalt and Coaldale before bending due north through Walker Lake, Fallon, and Lovelock. "The spreading center would connect with a transform fault coming in from Cape Mendocino," he adds, and he sketches such a line from the California coast to a point a little north of Lovelock. He is sketching the creation of a crustal plate, and he seems confident of that edge, for the Mendocino transform fault—the Mendocino trend—is in place now, ready to go. He is less certain about the southern edge of the new plate, because he has two choices. The Garlock Fault runs east-west just above Los Angeles, and that could become a side of the new plate; or the spreading center could continue south through the Mojave Desert and the Salton Sea to meet the Pacific Plate in the Gulf of California. "The Mojave sits in there with discontinued basin-and-range faulting," Deffeyes says, almost to himself, a substitute for whistling, as he sketches in the alternative lines. "There has to be a transform fault at the south end of the live, expanding rift. The sea has got to get through somewhere."

Now he places his hands on the map so that they frame the Garlock and Mendocino faults and hold between them a large piece of California—from Bakersfield to Redding, roughly, and including San Francisco, Sacramento, and Fresno—not to mention the whole of the High Sierra, Reno, and ten million acres of Nevada. "You create a California Plate," he says. "And the only question is: Is it this size, or the larger one? How much goes out to sea?" British Columbia is to his left and Mexico is to his right, beside his coffee cup on the oak Formica. The coast is against his belly. He moves his hands as if to pull all of central California out to sea. "Does this much go?" he says. "Or do the Mojave and Baja go with it?" A train of flatcars pounds through town carrying aircraft engines.

My mind has drifted outside the building. I am wondering what these people in this dry basin—a mile above sea level—would think if they knew what Deffeyes was doing, if they were confronted with the news that an ocean may open in their town. I will soon find out.

"What?"

"Are you stoned?"

"The way I see it, I won't be here, so the hell with it."

"It's a little doubtful. It could be, but it's a little doubtful."

"If it happens real quick, I guess a couple of people will die, but if it's like most other things they'll find out about it hundreds of years before and move people out of here. The whole world will probably go to hell before that happens anyway."

"You mean salt water, crests, troughs, big splash, and all that? Don't sweat it. You're safe here—as long as Pluto's out there."

"We got a boat."

"That's the best news I've heard in a couple of years. When I go bye-bye to the place below, why, that water will be there to cool me. I hope it's Saturday night. I won't have to take an extra bath."

"It may be a good thing, there's so many politicians; but they may get an extra boat. I used to be a miner. Oh, I've been all over. But now they've got machines and all the miners have died."

"The entire history of Nevada is one of plant life, animal life, and human life adapting to very difficult conditions. People here are the most individualistic you can find. As district attorney, I see examples of it every day. They want to live free from government interference. They don't fit into a structured way of life. This area was settled by people who shun progress. Their way of life would be totally unattractive to most, but they chose it. They have chosen conditions that would be considered intolerable elsewhere. So they would adapt, easily, to the strangest of situations."

"I've been here thirty-three years, almost half of that as mayor. I can't quite imagine the sea coming in—although most of us know that this was all under water at one time. I know there's quite a fault that runs to the east of us here. It may not be active. But it leaves a mark on your mind."

"Everybody's entitled to an opinion. Everybody's entitled to ask a question. If I didn't think your question was valid, I wouldn't have to answer you. I'd hope the fishing was good. I wouldn't mind having some beach-front property. If it was absolutely certified that it was going to happen, we should take steps to keep people out of the area. But as chief of police I'm not going to be alarmed."

"It'll be a change to have water here instead of desert. By God, we could use it. I say that as fire chief. We get seventy fire calls a year, which ain't much, but then we have to go a hundred miles to put out those damned ranch fires. We can't save much, but we can at least put out the heat. I got a ten-thousand-gallon tank there, which is really something for a place with no water. I guess I won't still be here to

see the ocean come, and I'm glad of it, because I can't swim."

Meanwhile, Deffeyes, in Sturgeon's Log Cabin, applies the last refining strokes to his sketchings on the map. "The Salton Sea and Death Valley are below sea level now, and the ocean would be there if it were not for pieces of this and that between," he says. "We are extending the continental crust here. It is exactly analogous to the East African Rift, the Red Sea, the Atlantic. California will be an island. It is just a matter of time."

Basin and Range

Lichen on rocks near Sundial Peak, Twin Peaks Wilderness, Utah.

6.   TAILINGS

WHAT DEFFEYES FINDS pleasant here in Pleasant Valley is the aromatic sage. Deffeyes grew up all over the West, his father a petroleum engineer, and he says without apparent irony that the smell of sagebrush is one of two odors that will unfailingly bring upon him an attack of nostalgia, the other being the scent of an oil refinery. Flash floods have caused boulders the size of human heads to come tumbling off the range. With alluvial materials of finer size, they have piled up in fans at the edge of the basin. ("The cloudburst is the dominant sculptor here.") The fans are unconsolidated. In time to come, they will pile up to such enormous thicknesses that they will sink deep and be heated and compressed to form conglomerate. Erosion, which provides the material to build the fans, is tearing down the mountains even as they rise. Mountains are not somehow created whole and subsequently worn away. They wear down as they come up, and these mountains have been rising and eroding in fairly even ratio for millions of years—rising and shedding sediment steadily through time, always the same, never the same, like row upon row of fountains. In the southern part of the province, in the Mojave, the ranges have stopped rising and are gradually wearing away. The Shadow Mountains. The Dead Mountains, Old Dad Mountains, Cowhole Mountains, Bullion, Mule, and Chocolate Mountains. They are inselberge now, buried ever deeper in their own waste. For the most part, though, the ranges are rising, and there can be no doubt of it here, hundreds of miles north of the Mojave, for we are looking at a new seismic scar that runs as far as we can see. It runs along the foot of the mountains, along the fault where the basin meets the range. From out in the valley, it looks like a long, buff-painted, essentially horizontal stripe. Up close, it is a gap in the vegetation, where plants growing side by side were suddenly separated by several metres, where, one October evening, the basin and the range—Pleasant Valley, Tobin Range—moved, all in an instant, apart. They jumped sixteen feet. The erosion rate at which the mountains were coming down was an inch a century. So in the mountains' contest with erosion they gained in one moment about twenty thousand years. These mountains do not rise like bread. They sit still for a long time and build up tension, and then suddenly jump. Passively, they are eroded for

Sage on the Wyoming plains.

North Schell Peak, Schell Creek Range of the Humboldt National Forest, Nevada.

millennia, and then they jump again. They have been doing this for about eight million years. This fault, which jumped in 1915, opened like a zipper far up the valley, and, exploding into the silence, tore along the mountain base for upward of twenty miles with a sound that suggested a runaway locomotive.

"This is the sort of place where you really do not put a nuclear plant," says Deffeyes. "There was other action in the neighborhood at the same time—in the Stillwater Range, the Sonoma Range, Pumpernickel Valley. Actually, this is not a particularly spectacular scarp. The lesson is that the whole thing—the whole Basin and Range, or most of it—is alive. The earth is moving. The faults are moving. There are hot springs all over the province. There are young volcanic rocks. Fault scars everywhere. The world is splitting open and coming apart. You see a sudden break in the sage like this and it says to you that a fault is there and a fault block is coming up. This is a gorgeous, fresh, young, active fault scarp. It's growing. The range is lifting up. This Nevada topography is what you see *during* mountain building. There are no foothills. It is all too young. It is live country. This is the tectonic, active, spreading, mountain-building world. To a nongeologist, it's just ranges, ranges, ranges."

Basin and Range

The Pfeifferhorn, Lone Peak Wilderness, Utah.

DEFFEYES AND I PASSED GRAFFITI ON the Bonneville flats. There being nothing to carve in and no medium substantial enough for sprayed paint, the graffitists had lugged cobbles out onto the hard mud—stones as big as grapefruit, ballast from the interstate—and in large dotted letters had written their names: ROSS, DAWN, DON, JUDY, MARK, MOON, ERIC, fifty or sixty miles of names. YARD SALE. Eric's lithography was in basalt and dolomite, pieces of Grayback Mountain, apparently, pieces of the Stansburys. His name, if it sits there a century or so, will eventually explode. Salt will work into the stones along the grain boundaries. When this happens, water evaporates out of the salt, and salt crystals keep collecting and expanding until they explode the rock. In Death Valley are thousands of little heaps of crumbs that were once granite boulders. Salt exploded them. Salt gets into fence posts and explodes them at the base.
Basin and Range

MORE MILES, AND THERE APPEARS ahead of us something like a Christmas tree alone in the night. It is Winnemucca, there being no other possibility. Neon looks good in Nevada. The tawdriness is refined out of it in so much wide black space. We drive on and on toward the glow of colors. It is still far away and it has not increased in size. We pass nothing. Deffeyes says, "On these roads, it's ten to the minus five that anyone will come along." The better part of an hour later, we come to the beginnings of the casino-flashing town. The news this year

Polished cobbles along the lake shore of Great Salt Lake.

is that dollar slot machines are outdrawing nickel slot machines for the first time, ever.
Basin and Range

THE HAIL OVER THE INTERSTATE turned to snow, and we passed a Consolidated Freightways tandem trailer lying off the shoulder with twenty-six wheels in the air—apparently overturned (a day or two before) by the wind. Abruptly, the weather changed, and we climbed the Rock Springs Uplift under blue-and-white marble skies. As we moved on to Green River and Evanston—across lake deposits and badlands, and up the western overthrust—the sun was with us to the end of Wyoming. On the state line was a flock of seagulls, in the slow lane, unperturbed, emblematically announcing Utah—these birds that saved the Mormons. Mormon traffic, heading home, did not seem intent on returning the favor.

If Wyoming can be said to have been acupunctured for energy, nowhere was this so variously evident as in the southwestern quadrant of the state, from the new coalfields near Rock Springs to the new oil fields of the Overthrust Belt, not to mention experimental attempts to extract petroleum from Eocene lacustrine shale, which—in that corner of Wyoming and adjacent parts of Colorado and Utah—contains more oil than all the rock of Saudi Arabia. More than the Union Pacific was after such provender now. "We are at the mercy of the East Coast and West Coast establishments," Love said. "It's been called energy colonization." And while we traversed the region, with scene after scene returning us to this theme, his reactions were not always predictable. There were moments that emphasized the scientist in him, others that brought out the fly-at-it-folks discoverer of resources, and others that brought forth a vigorous environmentalist, conserving his native ground, fulminating in the face of effronteries to humanity and the earth. Love is a prospector in the name of the people, who looks for the wealth in exploitable rock. He is also a pure scientist, who will follow his instincts wherever they lead. And he is a frequent public lecturer who turns over every honorarium he receives to organizations like the Teton Science School and *High Country News,* whose charter is to understand the environment in order to defend it. Thus, he carries within himself the whole spectrum of tensions that have accompanied the rise of the environmental movement. He carries within himself some of the central paradoxes of his time. Among environmentalists, he seems to me to be a good deal less lopsided than many, although beset by contradictory interests, like the society he serves. He cares passionately about Wyoming. It may be acupunctured for energy, but it is still Wyoming, and only words and images, in their inevitable concentration, can effectively clutter its space: a space so great that you can stand on a hilltop and see not only what Jim Bridger saw but also — through dimming tracts of time — what no one saw.

The Rock Springs Uplift, like the Rawlins Uplift, is a minor product of the Laramide Revolution, a hump in the terrain which did not keep rising as mountains. There was "red dog" — red clinker beds — in low cuts beside the road. When a patch of coal is ignited by lightning or by spontaneous combustion, it will oxidize the rock above it, turning it red. The sight of clinker is a sign of coal. Love said that this clinker was radioactive. Like coal, it was adept at picking up leached uranium. As the cuts became higher, we could see in the way they had been blasted the types of rock they contained. Where the cuts were nearly vertical, the rock was competent sandstone. Where the backslope angle was low, you knew you were looking at shale. Cuts that went up from the road through sandstone, then shale, then more sandstone, had the profiles of flying buttresses, firmly rising to their catch points, where they came to the natural ground. The shallower the slope, the softer the rock. The shallowest were streaked with coal.

At Point of Rocks, a hamlet from the stagecoach era, was a long roadcut forty metres high, exposing the massive sands of a big-river delta, built out from rising Rockies at the start of the Laramide Orogeny into the retreating sea. We left the interstate there and went north on a five-mile road with no outlet, which followed the flank of the Rock Springs Uplift and soon curved into a sweeping view: east over pastel buttes into the sheep country of the Great Divide Basin, and north to the white Wind Rivers over Steamboat Mountain and the Leucite Hills (magmatic flows and intrusions, of Pleistocene time), across sixty miles of barchan dunes, and, in the foreground — in isolation in the desert — the tallest building in Wyoming. This was Jim Bridger, a coal-fired steam electric plant, built in the middle nineteen-seventies, with a generating capacity of two million kilowatts — four times what is needed to meet the demands of Wyoming. Twenty-four stories high, the big building was more than twice as tall as the Federal Center in Cheyenne, which is higher than Wyoming's capitol dome. Rising beside the generating plant were four freestanding columnar chimneys so tall that they were obscured in cumulus from the cooling towers, which swirled and billowed and from time to time parted to reveal the summits of the chimneys, five hundred feet in the air. "This place is smoking the hell out of the country," Love said. "The wind blows a plume of corruption. In cold weather, sulphuric acid precipitates as a yellow cloud. It's not so good for people, or for vegetation. Whenever I think of this plant, I feel sadness and frustration. We could have got baseline data on air and water quality before the plant was built, and we muffed it." He blames himself, although at that time he had arsenic poisoning from springwater in the backcountry and was sick for many months.

Rising from the Plains

THIS STRIP MINE, NO LESS THAN AN erupting volcano, was a point in the world where geologic time and human time had intersected. Ordinarily, the close relationship between the two is masked: human time, full of beepers and board meetings, sirens and Senate caucuses, all happening in microtemporal units that physicists call picoseconds; geologic time, with its forty-six hundred million years, delivering a message that living creatures prefer to return unopened to the sender. In this place, though, geology had come up out of its depths to join the present world, and, as Love would put it, all hell had broken loose. "How people look at it depends on whose ox is being gored," he said. "If you're in a brownout, you think it's great. If you're downwind, you don't. Wyoming's ox is being gored."

When the Bridger operation was under construction, hundreds of tents and trailers lined most of the five miles of the spur road to the site—an "impact" that ultimately shifted to Rock Springs, thirty miles away, and Superior, and other small towns in the region. Populations doubled during the coal rush, which was close in time to the booms in trona mining and oil. Even after the booms had settled down, twenty-eight per cent of the people of Wyoming were living in mobile homes. During the construction of Jim Bridger, Rock Springs, especially, became a heavy-duty town, attracting people with no strong attachments elsewhere who came into the country in pickups painted with flames. With its bar fights and prostitutes, it was wild frontier territory, or seemed so to almost everyone but David Love. "Fights were once fights," he commented. "Now the fight starts and your friends hold you back while you throw insults." Cars were stripped of anything that would come off. Pushers arrived with every kind of substance that could stun the human brain. A McDonald's sprang up, of course, decorated with archaic rifles, with plastic cattle brands lighted from the inside, with romantic paintings of Western gunfights—horses rearing under blazing pistols on dusty streets lined with false-fronted stores. A Rock Springs policeman shot another Rock Springs policeman at point-blank range and later explained in court that he had sensed that his colleague was about to kill him. How was that again? The defendant said, "When a man has the urge to kill, you can see it in his eyes." The jury saw it that way, too. Not guilty. Some people in Sweetwater County seemed to be of the opinion that the dead policeman needed killing.

Love's son Charlie, who lives and teaches in Rock Springs, once told us that the community's underworld connections were "only at the hoodlum level." He explained, "The petty gangsters here aren't intelligent enough for the Mafia to want to contact. You can't make silk purses out of sows' ears."

The number of cowboys in Wyoming

**Flaming Gorge
National Recreation
Area, Utah.**

dropped from six thousand to four thousand as they rushed into town to join the boom, disregarding the needed ratio of one man per thousand head of cattle. In desperation for help at branding time, calving time, and haying time, ranchers had to go to the nearest oil rig and beg the roughnecks to moonlight.
Rising from the Plains

A FEW MILES SOUTH OF US WERE THE headwaters of the reservoir that covered Flaming Gorge. Before the federal Bureau

of Reclamation built a dam there, Flaming Gorge was a seven-hundred-foot canyon in arching Triassic red beds so bright they did indeed suggest flame. Afterward, not much was left but the hiss, and an eyebrow of rock above the water. The reservoir stilled fifty miles of river. Some of the high water penetrates beds of trona. When the reservoir drops, dissolved trona comes out of the rock and drips into the reservoir. When water rises again, it goes back into the rock for more trona. Love said that Lake Powell and Lake Mead —

reservoirs downstream—were turning into chemical lakes as a result. "And a lot of it winds up with the poor farmers in Mexico," he said. "We are going to have to desalinate their water." Some miles along the interstate, when we crossed the Blacks Fork River, we would see alkali deposits lying in the floodplain like dried white scum. On both sides of the road were abandoned farmhouses, abandoned barns, their darkly weathered boards warping away from empty structures out of plumb. The river precipitates and the abandoned farms were not unrelated. This was the Lyman irrigation project, Love explained—a conception of the Bureau of Reclamation, an attempt to make southwestern Wyoming competitive with Wisconsin. The Blacks Fork River was dammed in 1971, and its waters were used to soak the land. The land became whiter than a bleached femur. It still appeared to be covered with light snow. "Alkali sours the land," Love said. "The drainage here is just too poor to flush it out. Imagine the sodium those farmers drank in their water."

Meanwhile, west of Green River, a tall incongruous chimney seemed to rise up out of the range, streaming a white plume downwind. Below the chimney, but hidden by the roll of the land, was a trona refinery, and, below the refinery, a mine. I had gone down into it one winter day half a dozen months before, and I now remarked that the people there had told me that the white cloud issuing from the chimney was pure steam.

"It goes clear across the state," Love said. "That's pretty durable for steam."

He said that fluorine, among other things, was coming out of the refinery with the steam. Settling downwind, it could cause fluorosis. He thought it might be damaging forests in the Wind River

Range. The afternoon sky was cloudless but not exactly clear. "The haze you see is the trona haze that goes across Wyoming," he continued. "We never used to have this. You could clearly see distant mountains on any average day."
Rising from the Plains

HE SAID THAT SOMEWHERE IN LIMBO ON the industrial drawing board was a geothermal project that would mine the hot groundwaters of the Island Park Caldera, southwest of Yellowstone. The question uppermost in many people's minds seemed to be: What would happen to Old Faithful and other Yellowstone geysers? In New Zealand, when the government tapped the fifth-largest geyser field in the world for geothermal energy the Karapiti Blowhole shut down as promptly as if a hand had turned a valve. A geyser field in Nevada once rivalled Yellowstone's—until 1961, when geothermal well-drilling killed the Nevada geysers. Old Faithful was having trouble enough without help from the hand of man. For a century, and who knows how much longer, Old Faithful had erupted at intervals averaging seventy minutes, but in 1959 an earthquake centered nearby in Montana slowed the geyser down. Additional earthquakes in 1975 and 1983 caused Old Faithful to become so erratic that visitors complained. Constructed around the geyser is something that resembles a stadium, where crowds collect in bleachers and expect Old Faithful to be faithful: "to play," as hydrologists put it—to burst in timely fashion from its fissures, like a cuckoo clock made of water and steam. Frustrated travellers, sometimes clapping their hands in unison, seemed to be calling on the National Park Service to repair the geyser. A scientist confronted with these facts could only

shrug, make observations, and formulate a law: *The volume of the complaints varies inversely with the number of miles per gallon attained by the vehicles that bring people to the park.*

Love, who has made a subspecialty of the medical effects of geology, had other matters on his mind. In public lectures and in meetings with United States senators, he asked what consideration was being given to radioactive water from geothermal wells, which would be released into the Snake River through Henrys Fork and carried a thousand miles downstream. After all, radioactive water was known from Crawfish Creek, Polecat Creek, and Huckleberry Hot Springs, not to mention the Pitchstone Plateau. On the Pitchstone Plateau were colonies of radioactive plants, and radioactive animals that had eaten the plants: gophers, mice, and squirrels with so much radium in them that their bodies could be placed on photographic paper and they would take their own pictures. A senator answered the question, saying, "No one has brought that up."
Rising from the Plains

IN THE BRONCO, WE MOVED THROUGH the snow toward the mountains, crossing the last of the Great Plains, which had been shaped like ocean swells by eastbound streams. Now and again, a pump jack was visible near the road, sucking up oil from deep Cretaceous sand, bobbing solemnly at its task—a giant grasshopper absorbed in its devotions. As we passed Cheyenne, absolutely all we could see in the whiteout was a raging, wind-whipped

Upper Geyser Basin, Yellowstone National Park.

flame, two hundred feet in the air, at the top of a refinery tower. "Such a waste," said Love.
Rising from the Plains

DRAFTING HIS REPORT TO THE GEOLOGical Survey, Love described the "soft porous, pink or tan concretionary sandstone rolls in which the uranium was discovered," and added that "the commercial grade of some of the ore, the easy accessibility throughout the area, the soft character of the host rocks and associated strata, and the fact that strip-mining methods can be applied to all the deposits known at the present time, make the area attractive for exploitation." With those sentences he had become, in both a specific and a general sense, the discoverer of uranium in commercial quantity in Wyoming and the progenitor of the Wyoming uranium industry—facts that were not at once apparent. Within the Survey, the initial effect of Love's published report was to irritate many of his colleagues who were committed hydrothermalists and were prepared not to believe that uranium deposits could occur in any other way. They were joined in this opinion by the director of the Division of Raw Materials of the United States Atomic Energy Commission. A committee was convened in the Powder River Basin to confirm or deny the suspect discovery. All the members but one were hydrothermalists, and the committee report said, "It is true that high-grade mineral specimens of uranium ore were found, but there is nothing of any economic significance." Within weeks, mines began to open in that part of the Powder River Basin. Eventually, there were sixty-four, the largest of which was Exxon's Highland Mine. They operated for thirty-two years. They had removed fifteen million tons of uranium ore when

Three Mile Island shut them down.

In 1952, after Love's report was published, the *Laramie Republican and Boomerang* proclaimed in a banner headline, "LARAMIE MAN DISCOVERS URANIUM ORE IN STATE." The announcement set off what Love described as "the first and wildest" of Wyoming's uranium booms. "Hundreds came to Laramie," he continued. "I was offered a million dollars cash and the presidency of a company to leave the U.S.G.S. At that time, my salary was $8,640.19 per year."

The discovery predicted uranium in other sedimentary basins, and Love went on to find it. In the autumn of 1953, he and two amateurs, all working independently, found uranium in the Gas Hills—in the Wind River Basin, twelve miles from Love Ranch. By his description: "Gas Hills attracted everybody and his dog. It was Mecca for weekend prospectors. They swarmed like maggots on a carcass. There was claim-jumping. There were fistfights, shootouts. Mechanics and clothing salesmen were instant millionaires."

As it happened, he made those remarks one summery afternoon on the crest of the Gas Hills, where fifty open-pit uranium mines were round about us, and in the low middle ground of the view to the north were Muskrat Creek and Love Ranch. The pits were roughly circular, generally half a mile in diameter, and five hundred feet deep. Some four hundred feet of overburden had been stripped off to get down to the ore horizons. The place was an unearthly mess. War damage could not look worse, and in a sense that is what it was.

"If you had to do this with a pick and shovel, it would take you quite a while," Love said. The pits were scattered across a hundred square miles.

We picked up some sooty black uraninite. It crumbled easily in the hand. I asked him if it was dangerously radioactive.

"What is 'dangerously radioactive'?" he said. "We have no real standards. We don't know. All I can say is the cancer rate here is very high. There are four synergistic elements in the Gas Hills: uranium, molybdenum, selenium, and arsenic. They are more toxic together than individually. You can't just cover the tailings and forget about it. Those things are bad for the environment. They get into groundwater, surface water. The mines are below the water table, so they're pumping water from the uranium horizon to the surface. There has been a seven-hundred-per-cent increase of uranium in Muskrat Creek at our ranch."

We could see in a sweeping glance—from the ranch southwest to Green Mountain—the whole of the route he had taken as a boy to cut pine and cedar for corral poles and fence posts. An hour before, we had looked in at the ranch, where most of those posts were still in use—gnarled and twisted, but standing and not rotted. From John Love's early years there, when he slept in a cutbank of the creek, the ranch had belonged only to him and his family. The land was leased now—as was most of the surrounding range—to cattle companies. In the last half mile before we reached the creek, David counted fifty Hereford bulls and remarked that the lessees seemed to be overgrazing. "The sons of bitches," he said. "That's way too many for this time of year." Noticing some uranium claim stakes, he said, "People stake illegally right over land that has been deeded nearly a century."

Mine tailings of copper mines near Ely, Nevada.

Over the low and widespread house, John Love's multilaminate roof was scarcely sagging. No one had lived there in nearly forty years. The bookcases and the rolltop desk had been removed by thieves, who had destroyed doorframes to get them out. The kitchen doorframe was intact, and nailed there still was the board that showed John Love's marks recording his children's height. The green-figured wallpaper that had been hung by the cowboys was long since totally gone, and much of what it had covered, but between the studs and against the pine siding were fragments of the newspapers pasted there as insulation.

POSSE AFTER FIVE BANDITS
BATTLE NEAR ROCK ISLAND TRAIN

Robbers Are Found in Haystack and
Chase Becomes Hot

BOTH SIDES ARE HEAVILY ARMED

Fugitives Are Desperate, but Running
Fight Is Expected to End in Their Capture

Spinach had run wild in the yard. In the blacksmith shop, the forge and the anvil were gone. Ducks flew up from the creek. There were dead English currant bushes. A Chinese elm was dead. A Russian olive was still alive. David had planted a number of these trees. There was a balm of Gilead broadleaf cottonwood he had planted when he was eleven years old.

"It's going to make it for another year anyway," he said. "It's going to leaf out."

I said I wondered why the only trees anywhere were those that he and his father had planted.

"Not enough moisture," he replied. "Trees never have grown here."

"What does 'never' mean?" I asked him.

He said, "The last ten thousand years."

An antelope, barking at us, sounded like a bullfrog. Of the dozen or so ranch buildings, some were missing and some were breaking down. The corrals had collapsed. The bunkhouse was gone. The cottonwood-log granary was gone, but not Joe Lacey's Muskrat Saloon, which the Loves had used for storing hay. Its door was swinging in the wind. David found a plank and firmly propped the door shut. The freight wagon was there that he had used on trips for wood. It was missing its wheels, stolen as souvenirs of the Old West. We looked into a storage cellar that was covered with sod above hand-hewn eighteen-inch beams. He said that nothing ever froze in there and food stayed cold all summer. More recently, a mountain lion had lived there, but the cellar was vacant now.

In the house, while I became further absorbed by the insulation against the walls, Love walked silently from room to room.

Bizerta, Tunis, May 4—At a reception tendered him by the municipality, M. Pelletan, French Minister of Marine, in a brief speech, declared that France no longer dreamed of conquests, and that her resources would hereafter be employed to fortify her present possessions.

Cattle chips and coyote scat were everywhere on the floors. The clothes cupboards and toy cupboards in the bedroom he had shared with Allan were two feet deep in pack-rat debris.

Have you lost a friend or relative in the Klondike or Alaska? If so, write to us and we will find them, quietly and quickly. Pri-

Dead cottonwoods.

Winter dawn, Grand Teton National Park, Wyoming.

vate information on all subjects. All correspondence strictly confidential. Enclose $1.00. Address the Klondike Information Bureau, Box 727, Dawson, Y.T.

David came back into the space that had been his schoolroom, saying, "I can't stand this. Let's get out of here."

In the Gas Hills, as we traced with our eyes his journeys to Green Mountain, he said, "You can see it was quite a trek by wagon. Am I troubled? Yes. At places like this, we thought we were doing a great service to the nation. In hindsight, we do not know if we were performing a service or a disservice. Sometimes I think I might regret it. Yes. It's close to home."

Rising from the Plains